Gender, Caste, and Class in South India's Technical Institutions

Through its engaging ethnographic writing and sharp theorization of questions of gender, caste, and class, the book is an insightful study of youth culture and student aspirations in a private engineering college in Tamil Nadu and constitutes an important contribution to the sociology of post-liberalization India. Hebbar skilfully uncovers the ways in which young women students negotiate with gendered norms of 'respectability' imposed by their families, communities, and the college itself, and portrays with sensitivity their hopes, dilemmas, and disappointments.

Carol Upadhya, Honorary Visiting Professor, National Institute of Advanced Studies, Bengaluru

Nandini Hebbar's book is a fascinating exploration into the world of private engineering education in South India. Based on intensive fieldwork, it unveils the connections between the interests of 'big men' who control such education and the social mobility desires of families that drive gendered investments in children, the aspirations of girls, and the role of gender, caste, and class in shaping their educational outcomes.

Ravinder Kaur, Emerita Professor, Department of Humanities and Social Sciences, Indian Institute of Technology Delhi

Through a study of young women in an engineering college in Tamil Nadu, Nandini Hebbar weaves together a range of critical issues to tell a story about contemporary India—about caste and higher education, the aspirations of young women, of caste identities and pride, of familial anxieties, restrictions, and control. Written in a highly readable and clear style, this rich ethnography should be taught in undergraduate and postgraduate social science programmes in India and abroad.

Janaki Abraham, Professor, Department of Sociology, Delhi School of Economics, University of Delhi

EDUCATION AND SOCIETY IN SOUTH ASIA
SERIES EDITOR
Meenakshi Thapan

A series on Education and Society in South Asia is, perhaps, somewhat ambitious given the magnitude and scope of the theme in present times. It is hoped that a broad framework will allow the inclusion of work that seeks to understand education and society in both institutional and non-institutional settings. Institutional frameworks bring a focus on schools and institutions for higher education as spaces within which educational processes and activities unfold. Non-institutional spaces include all those other social, cultural, and political spaces where pedagogic encounters of different kinds take place. They also include movements, trajectories, and patterns of different kinds that enable children and youth to learn and grow in unconventional ways. One of the lenses through which this series aims to develop the sociology of education is the focus on children and youth and aspects of their experience both within and outside institutional spaces.

At one level, this series seeks to problematize our understanding of education, as process, in the context of the making of citizens in a 'modern', changing South Asia. Most efforts view schools, for example, as institutions that transmit and evaluate educational knowledge and provide certification based on academic achievement. The causes of inequality, located in gender, caste, class, and religion have been examined in this context, as these shape the lives of individuals in multiple and complex ways. At the same time, schools and institutions for tertiary education are spaces, as processes, through which participants bring meaning and create worlds that hugely impact their personal and intellectual development.

Educational spaces are also about place and location in multiple ways, whether these are at the intersection of caste, gender, and class or are about location as both territorial and imagined spaces in the region. The sociology of education must unpack these complexities and bring out their implications in a variety of contexts in both rural and urban South Asia. The significance of gender, caste, religion, or language as defining characteristics of educational processes are germane to our understanding and need to be examined in different contexts in the region that make each experience unique and similar at the same time.

Conflict, crises, and events in everyday life are significant aspects of these processes. The ways in which youth may be both shaped by as well as engage with the unfolding of crises, events, and everyday life remain opaque in our understanding of how education plays an important role in the making of citizens in contemporary South Asia. It is this understanding of human agency in institutional contexts that has somehow eluded those arguments that over-emphasize the significance of the structural and ideological frameworks within which educational processes are embedded. Once we understand that students are keen and active participants in the processes in which they are inserted, our views about education and its possible outcomes may perhaps change. It is indeed possible to examine and understand the vastly differing and multiple practices that students engage in as agents within and outside an institutional framework. We need to, for example, focus on and unravel the complexities prevalent in understanding youth as active citizens in an increasingly cosmopolitan and interconnected world. How do they seek to rise above not just the normative expectations associated with their 'roles' as students but also with asserting themselves in deeply meaningful and contextually significant ways? This means that we must pay attention to critical consciousness as it reveals itself in pedagogic encounters of different kinds but also in peer cultures and student-led organizations and movements in different parts of South Asia.

The goal is to sharpen our understanding of perspectives that do not rely on educational misgivings, institutional features, financial outlays, and state failures alone. These are only some aspects of educational practice. There are other dimensions that envelop students, teachers, the community, and society in complex ways that we need to uncover in order to provide an understanding of how education and society connect in diverse ways.

Teachers are agents of both change as well as reproduction in education. It is important to identify and examine some of the processes that enable them to be pioneers and facilitators for transformative practices rather than only being viewed as toothless agents of the state or private bodies, as they usually are, without any possibilities for bringing about change. Both teachers and students are engaged participants in processes pertaining to education, and the series must unpack the possibilities and potential for movement underlying the constrained and encapsulated worlds they inhabit. In this context, a focus on mainstream educational processes that incorporate a social, ecological, and moral vision for inclusive education is an endeavour for the series. This will allow an analysis of practices that focus on the well-being and holistic growth of all stakeholders in a changing and globalizing society.

Above all, the series is an initiative in the discipline of sociology wherein research on education is a somewhat neglected dimension of the broader disciplinary framework in India and the rest of South Asia. As Basil Bernstein, late Karl Mannheim Professor of Sociology of Education at the University of London Institute of Education, used to tell me: 'My dear, we're at the bottom of the pile! No one in sociology wants to study education'. Almost forty years later, with some improvement in educational studies in sociology, this series seeks to redress this lacuna by focusing precisely on a disciplinary area that begs attention.

Meenakshi Thapan is Director of the Rishi Valley Education Centre (Krishnamurti Foundation India) in rural Andhra Pradesh. Previously she was Professor of Sociology and Director, Delhi School of Economics, University of Delhi, India.

Gender, Caste, and Class in South India's Technical Institutions

NANDINI HEBBAR N.

OXFORD
UNIVERSITY PRESS

Great Clarendon Street, Oxford, OX2 6DP,
United Kingdom

Oxford University Press is a department of the University of Oxford.
It furthers the University's objective of excellence in research, scholarship,
and education by publishing worldwide. Oxford is a registered trade mark of
Oxford University Press in the UK and in certain other countries

© Nandini Hebbar N. 2024

The moral rights of the author have been asserted

All rights reserved. No part of this publication may be reproduced, stored in
a retrieval system, or transmitted, in any form or by any means, without the
prior permission in writing of Oxford University Press, or as expressly permitted
by law, by licence or under terms agreed with the appropriate reprographics
rights organization. Enquiries concerning reproduction outside the scope of the
above should be sent to the Rights Department, Oxford University Press, at the
address above

You must not circulate this work in any other form
and you must impose this same condition on any acquirer

Published in the United States of America by Oxford University Press
198 Madison Avenue, New York, NY 10016, United States of America

British Library Cataloguing in Publication Data

Data available

Library of Congress Control Number: 2024936473

ISBN 978-0-19-891445-7

DOI: 10.1093/9780198914488.001.0001

Printed and bound in India by
Replika Press Pvt. Ltd.

Dedicated to the memory of my Ajja

Contents

Figures   xi
Tables   xiii

Introduction: The Engineering Mania   1

1. 'Kinning' Education: Genealogies of Private Engineering Colleges   47
2. 'Edupreneurship': Mapping Management Practices   77
3. Becoming Professional: Dilemmas in Emerging 'Employable'   109
4. Manufacturing Respectability: Gendering the Engineering College Boom   141
5. Negotiating Intimate Risk: Gendered Subjectivities, Performativity, and Self-Making   181
6. Engineering Aspirations and Lives of Youth: Implications for Gender, Caste, and Class   215

Bibliography   233
Index   255

Figures

4.1. Average number of male and female students per year in various streams in 2014 — 150

4.2. The T-shirt designed by students of the Mechanical Engineering Class for their department festival — 157

Tables

3.1.	First Graduate by Stream	112
3.2.	First Graduate by Category and Gender	112
3.3.	Management Quota by Category and Gender	113

Introduction
The Engineering Mania

I am in western Tamil Nadu, taking a noisy rattling bus from Salem City to the neighboring district of Namakkal. It is a crowded 'red bus'—a moniker given to state-run cheap transport plying the hinterlands—and people are even seated on the floor. I am at the front end of the bus, crammed into a bench-seat adjacent to the engine. As we make our way out of the city, I can see a school bus in front of us advertising its host institution with the tagline 'Training for future engineers'. I take it as a sign that my work has begun well.

However, the excitement is soon replaced by nervousness. As we leave the city behind, I start fretting about where to get off. The friend who had dropped me to the bus station had asked me to tell the driver to drop me in front of M. College of Engineering on the highway.

'There is no stop?! What if I miss it?' I had asked.

'You won't miss it', she had said.

Given her confidence, I try not to worry, and look out of the window so as to not miss a single sign. We pass fields of maize and groundnut, areca gardens, and in clearings, I see a couple of Aiyyannar temples with terracotta figures of mustachioed heroes and horses. The area is semi-arid and the owners of the agricultural farms here have been incredible resourceful in diversifying their interests: trucks, automobile parts, livestock, and other small businesses.

This is also the famed Poultry Corridor of Tamil Nadu: home to egg-laying poultry farms as much as the educational institutions I want to write about. Some critics have gone as far as to suggest that many similarities mark the two: unbridled growth, home-grown feeder mechanisms, automated systems aimed at efficiency, and layered cages! Even if this analogy is intended to be tongue-in-cheek, the success of poultry farms has played an important role in securing capital for setting up engineering colleges in this region, along with other businesses such as truck and lorry services and manufacturing spare parts for the automobile industry. These businesses are impossible to miss as we cross over to Namakkal district.

About thirty minutes pass, and I am busy taking in the landscape. By now, billboards by the side of the road are advertising M. College of Engineering, and I guess that we must be getting close to my destination. I have an appointment to meet a top-ranking administrator at M. College to explore the possibility of staying on campus for my research. I don't want to miss my stop, so I collect my things and begin making my way to the exit. Another ten minutes pass. I am sorry that I have let my seat go as I struggle to retain my balance in the swinging bus. Finally, we come to an arch by the side of the road, with tall wrought iron gates. The bus driver signals me to get off, but when I peek outside, the board says that this is the M. Arts and Science College, so I tell him that this is not it, and stay on.

We hurtle forward on the road. I have my eyes glued to the high compound wall and barbed wire that run parallel to the road, marking the campus perimeter. Another ten minutes pass, but the wall continues. The campus is much bigger than I imagined! Finally, the bus draws close to another large gated archway. This is my stop. I get off and approach the gate. A security guard stops me.

'Enge poganumm?' (Where do you want to go?)

I tell him I am here to meet Mr IR, the Education Director of the institution.

'Oh, ED Sir', he says, scribbling it on a pink slip.

Meanwhile, he instructs me to enter my name and details in a register lying on the counter. The time is 11.45 am. I am just in time for the interview scheduled at 12 noon. He hands me the pink slip with 'Visitor's Pass' in bold. He asks me to walk straight ahead to the administrative block.

With the slip in hand, I turn to take in my surroundings. The driveway is divided into two lanes with broad pavements on either side. I start walking. The sun beats down. I have to remind myself it is a college campus for it is eerily quiet. 'Maybe it's vacation time', I think to myself. I spot a few students closer to the buildings, studying, but there are no groups of friends 'hanging out' or walking in and out of campus.

The driveway is long, and the building at the far end is imposing. It has a façade of glass and chrome, pistachio-green tints, combined with features of the 'neo-Dravidian' architectural style—an idiom common to temples, but increasingly used for secular buildings such as hospitals and colleges in Tamil Nadu. As if to make the connection explicit, the words 'Temple of Learning' are also displayed prominently across the building. At the reception, I ask to see Mr IR. The receptionist nods and picks up the phone, speaks a few words before hanging up. He asks me to follow him. We pass a few classrooms and labs. The classrooms have walls of glass, so anyone walking on the corridor can easily see inside.

'Where are the students?' I asked the receptionist.

He replies that the semester exams are in progress, engulfing the campus in an anxious silence.

We approach a door on which Mr IR's name is displayed prominently with the title 'Education Director' underneath. I am shown into Mr IR's huge chambers, with a writing desk as well as a conference table. I take a seat and describe to him the objectives of the participant observation research that I would like to conduct in the college. He agrees to forward my proposal to the college chairman and asks if he can order lunch for me to eat in his chambers. I politely decline saying that I would prefer to visit the canteen instead. He offers to drop me at the 'food court', and invites me into his gleaming black Sports Utility Vehicle (SUV) parked in the front. After a short drive, the vehicle drops me off in front of a diminutive building, painted dark green with a soft drink advertisement hanging over the doorway. This is the college food court. To one side, along the length of the room, there is a food counter. On the other side, there are tables and chairs on which sit many young men, eating and chatting with their friends.

'Why are there only men here?' I wonder, as I go through the menu displayed over the counter. While I am trying to decide what to order, I cannot shake off the feeling that everyone is staring at me. I try to be nonchalant and decide on having sambharsaadam (pre-mixed sambhar and rice) for lunch. I approach the counter to order and see many young women on the other side of the kitchen. That is when I finally realise why I am being stared at—I am in the men's section! I had mistaken the food counter to be at one end of the room, when actually the long room is bisected into separate men and women's sections! I walk out of the building and enter through a second entrance into the women's section.

I had been initiated into the world of private engineering colleges in Tamil Nadu.

Field Journal, Salem/Namakkal, 5 December 2013

Landscapes across India are distinctively marked by the visibility given to professional courses such as engineering, management, and medicine in the form of massive construction for colleges and billboards promoting institutions. Of these paths, a career in engineering emerged as the most desirable in the 1990s and early 2000s, partly because of the Information and Communication Technology (ICT or IT) boom, and the consequent rise in opportunities for engineering graduates.[1] The advantages of a professional degree combined with the chance of a lucrative career in a multinational company amplified the prestige associated with engineers since colonial times (Ramnath 2017). As the demand grew, so did the number of private engineering colleges. Efforts at privatization of education, which had begun in the 1970s and 1980s, through the loosening of various regulations[2] (Kaul 1993; Pinto 1994), achieved fruition through the IT boom, when the number of colleges, especially offering courses in engineering and management, increased manifold.

According to the All India Council of Technical Education (AICTE) website, the apical regulatory body of engineering education, there are 2901 undergraduate engineering colleges in India (as of 2023–2024), with a total intake of over 1.35 million. Of all undergraduates enrolled in higher education, 16.05 per cent are in engineering—making it the most sought-after professional course, and the fourth-most popular degree after a BA, a BCom and a BSc, which are offered mostly by colleges under

---

[1] For software companies, the recruitment of engineering graduates as IT workers promised a workforce fulfilling requirements such as the technical competency required to learn coding and programming as well as English and communication skills. 'Engineers' were also more likely to be granted business visas for when they had to travel to the US or Europe to work 'on-site.' Moreover, engineering institutes were known to attract the best talent, given the durability of the Nehruvian vision of India as a great scientific technocracy. Companies were lured by the 'merit' of students from prestigious state-funded schools such as the Indian Institute of Technology (IIT) as well as private engineering colleges known to churn out thousands of engineering graduates every year. Such a preference for engineers only amplified the existing status of the engineering degree in India, especially among the middle classes keen to 'professionalize' in the absence of other forms of symbolic and social capital (Beteille 1991; Wilson 2008; Ramnath 2017).

[2] The AICTE, which had existed as a regulatory body since 1945, received a shot in the arm through an Act of Parliament in 1987 allowing private and voluntary organizations to start colleges on a self-financing basis.

public-sector funding (AISHE 2019–20, II). Undergraduate engineering, on the other hand, is almost completely in the realm of private unaided institutions, with 69.85 per cent share of total places (AICTE website n.d.). The five South Indian states and Maharashtra account for more than half the total number of engineering colleges in the country. In Tamil Nadu, where the research for this book is based, the number of colleges went from ten in the early 1980s to a whopping 514 private self-financed institutions in 2015–2016 (AICTE website). Though this number has seen a decline; about seventy-eight engineering colleges in Tamil Nadu and 556 colleges in India have closed since (AICTE website n.d.), the boom has been responsible for crafting a distinctive middle-class culture associated with privatized education. This explosion of private institutions has not come from any change in ideological commitments of the state and its regulatory bodies, which still categorize higher education as a public good deserving state investment. Instead, it has sprung from a failure in governance which has practically delivered education to ideological and vested interests (Kapur and Mehta 2004). This has widespread implications ranging from the usurping of philanthropy by private bodies to middle-class loss over control of public-sector education. Though couched in the language of the 'success' of private education, this book echoes Kapur and Mehta (2004) in saying that the result is actually a crisis of the widest proportions: relating to the public use and mobilization of private resources, unemployment, reproduction of gendered and casted forms of sociality, as well the construction of 'technical subjects' not only through the form and mode of education, but also as a form of legitimate global citizenship.

This book attempts to make sense of the processes of privatization as well as the aspirations, strategies, and crises outlined above through an ethnographic study of engineering colleges in Tamil Nadu, one of the states in which the engineering college boom is most visible. As in many other South Indian states, colleges in Tamil Nadu are run by 'institutional' big men, often in ways that intersect with other commercial interests, patronage projects, and electoral politics. The architecture described in the prologue, along with features such as gender segregation, strict disciplinary positions, surveillance mechanisms, and the use of a neoliberal idiom to refer to campus spaces have become an inextricable part of the aspirational culture authored and sustained by these colleges.

Both office- and temple-like in appearance, these colleges are veritable artefacts of the 'engineer worship' (Ramnath 2017) that characterizes educational cultures in India today[3]. The objective of this book is to map the historical antecedents of that culture, the contemporary contexts and crises created by it, as well as to document the diverse lived experiences in those spaces.

Intensive ethnographic fieldwork for the book was conducted over ten months in 2014, when I lived among students in the hostel of an engineering college and attended classes with them. However, engagement with engineering colleges, their histories, organizational structures, their management practices, trends in enrolment and curriculum, as well as the relationships cultivated with interlocutors on the field continued for about five or six years after fieldwork, giving this book a longitudinal perspective that transcends an ethnographic study of student life.

The Contemporary Contexts of Engineering Education

Though engineering education has a long history linked to colonial imperialism and is currently inextricable from the global spike in knowledge work, this book integrates a sociocultural perspective while examining historical and political economy changes to look at contemporary contexts of education through its linkages with regional histories and local caste–class politics. Thus, even though certain terms discussed in this book such as 'privatization', 'merit', 'employability', and 'dress codes' are the subject of nationwide debates, I place them within an economy of everyday exchanges, desires, aspirations, and anxieties of life in college to understand the ways in which they are deployed in the 'field'. This not only portrays the culture of aspiration, but also shows the ways in which engineering colleges are enmeshed in the local fabric, despite their academic orientation to creating a workforce for the offshore economy. The

---

[3] Though inferences drawn from this book may be extended to private colleges offering different kinds of professional colleges (such as, but not limited to],medical, dental, architecture, and management or all of them as in the case of some 'educational empires'), this book limits itself to engineering colleges to which the proliferation argument is most applicable for reasons explained earlier. Similarly, even though the arguments are made in reference to private colleges in Tamil Nadu due to the limits posed by my fieldwork, they might be applicable to other regions, especially in the south.

local identities of the college, based on factors such as caste of the college management or industrial/symbolic lineages, mark the colleges as being more than mere 'training centres' for the creation of 'borderless entrepreneurial subjects', as Aihwa Ong (2006, 149) fears has happened with the globalization of tech schools in the Information Age. They are agents of their own history in the way they aspire, strategize, and attempt to align their trajectories with global trends of capital even as they adhere to local norms and secure their own ends and objectives.

Despite India's interpellation in the knowledge economy as a key producer of software goods for countries like the US (Saith and Vijayabaskar 2005; Vasavi and Upadhya 2006; Upadhya 2016b)—an identity that has changed the way the country projects itself to the world (Varma 2004), and the nation's youth imagine their futures (Poonam 2018), the career paths, and social mobilities charted are along beaten tracks. As Carol Upadhya argues in her book *Reengineering India* (2016b), despite becoming one of the prime destinations for knowledge work, the New India is very much embedded in older forms of class and capital. The hundreds of colleges that have sprung up to cater to the rise in IT work pivot around older modalities and forms of higher education, making few changes to adapt to what Upadhya (2016b) calls the 'novelness' of IT work.[4] However, apart from a few in-depth studies in the 1990s on the then 'new phenomenon' of 'capitation fee colleges' (for instance Kaul 1993; Upadhya 1997), scholarly literature on higher education in general, and private education in particular, has been sparse. Even in this literature, engineering colleges are relegated to a footnote when not dealing with elite institutions such as the Indian Institutes of Technology. Of course, many have put forth very powerful critiques of the governance and machinery of education that remains committed, in principle, to state control over education, even as it has allowed the private system to flourish in the name of pragmatism and adaptation to changing economic climes. Devesh Kapur and Pratap Bhanu Mehta (2004) evocatively refer to the resultant unevenness in the system as the transition from 'half-baked socialism' to 'half-baked capitalism'. This shift was espoused by political parties across the spectrum in

---

[4] Novel in terms of how forms of work, worker subject positions, aspirations, and ways of balancing work and family are visualized. Also see Smitha Radhakrishnan (2011) and Jyothsna Belliappa (2013).

the 1990s (Kaul 1993; Pinto 1994; Upadhya 1997; Lukose 2009) as part of a consensus that state coffers could not afford to build and maintain a robust public education system to keep up with the demand. Moreover, the increase in supply of seats was also perceived to be the end of the reservation conundrum and facilitate new forms of patronage as older forms get exhausted (Kapur and Mehta 2004, 7). As a result of this, almost all states enacted regulations for private players to enter the sphere of higher education, sometimes even exploiting loopholes in the central regulations to enable entry.

However, no in-depth study exists on the effects of privatization—the role and models of patronage, new caste/class compositions in the classrooms, its effects not just in the everyday, but also on questions of employment and marriage. Indian social science academia's own elite locations in higher education perhaps translated into a bias against private institutions, especially technical schools that run on capitation fees. Even the recent crop of youth studies based in and around Tamil Nadu (Lukose 2009; Rogers 2008; Jeffrey 2010; Nakassis 2016; Krishnan 2014) do not address engineering colleges, which have seen the biggest spike since privatization, despite dealing with increased backward class presence on campuses. In fact, as Constantine Nakassis (2013) notes, his argument of youth being exterior to the standard and normative rules of society would not withstand universal application in colleges in Tamil Nadu—they would not hold true for young women or under disciplinary conditions enforced in private colleges where efforts are specifically made to stake authority on the 'exteriority' carved by youth.

This book seeks to address this empirical lacuna in the scholarship on higher education in India, by attempting to understand different facets of private higher education, from funding to business models, questions of employment to student subjectivities. As compared to state-funded schools that are governed according to centralized norms, private engineering colleges are allowed many liberties as long as a few basic mandates set up by the AICTE are met,[5] making them akin to private fiefdoms of the college managements.

---

[5] The prerequisites to start a college are set as follows: 1.5 acres (7260 sq. m.) of land in an urban settingor 7.5 acres (36,300 sq. yards.) in a rural setting, classrooms having a minimum of 66 sq. ft. carpet area, and a crore (roughly $1.2 million US) as operational expenditure. See Nanda (2016).

In Tamil Nadu, though sixty-five per cent of college places are surrendered to the government for centralized admissions ('government quota'). Colleges are allowed to decide their own criteria for admission to the other thirty-five per cent ('management quota'), recruit their own staff, and enforce student codes of conduct as they deem suitable. Places in the management quota are notorious for being money spinners, as they can be bought through payment of a donation (also called management fees or capitation fees) to the college trust. Even as academic elites have criticized the 'capitation fee colleges' (Pinto 1994), the model is hugely popular and thousands throng to apply for places through the management quota in a college of their choice, rather than wait for allotment of places by the government. It is a popular perspective that management quotas provide a means of bypassing the 'undemocratic' and 'political' public universities (Lukose 2009) with their tedious bureaucratic procedures, adherence to reservation quotas, and 'merit lists'. By paying a fee, students can secure admission in a college of their family's choice based on factors such as college reputation, location, convenience, finances, and community affiliation of the college management. Investments in education from groups that were historically disadvantaged but have developed the financial capability to pay for higher education are also 'quests for inclusion' (Ong 2006; Shklar 1991). 'Buying a seat' represents a means of showcasing clan mobility, a *consumption choice* that generates 'recognition' and 'visibility' and thereby middle-class status (Dickey 2013). These factors have been instrumental in driving the expansion of education and its commercialization and feeds the popular narrative on privatized higher education as providing greater 'scope' for individual and inter-generational mobility (Lukose 2009, 155–162).

Without a doubt, privatization has ensured greater enrolment among hitherto unrepresented or underrepresented groups in higher education, particularly in streams like engineering which used to be a bastion of the elite (Subramanian 2015a; Fuller and Narasimhan 2007). For instance, there is unprecedented enrolment of women in higher education (John 2019) with a significant number in the STEM (Science, Technology, Engineering, and Maths) disciplines. According to the AICTE, the male-to-female ratio of engineering students at the undergraduate level in 1981 was 0.04, rose to 0.08 by 1986, to 0.28 by 2001, to 0.5 in 2012, and currently stands at 0.4 (Nair 2012, 8; AICTE Enrolments: Gender and Category Wise on website, n.d.). This is also largely because of the rise of streams such as software and information technology, which are seen as 'decent' and 'respectable' for middle-class young women to pursue. Similarly, more than thirty per cent of students in engineering colleges are the first-ever graduates in the family and more than two thirds belong to groups such as 'Other Backward Class', 'Most Backward Class', Scheduled Castes, and Scheduled Tribes. The state has also integrated social welfare measures into private higher education, mandating colleges to follow reservation policies as well as paying fees on behalf of students who are from Scheduled Caste, Scheduled Tribe groups, or the first-ever graduates in their families.

However, as Carol Upadhya (2016a) points out, state discourses of education leading to upward mobility tend to invisibilize the role played by historical and regional formations of capital, class, and caste in producing 'success stories' (Upadhya 2016a; also Kamat 2011). This invisibility misleadingly portrays the road to success as a straightforward, merit-oriented process in which credentialed qualifications result in well-paying jobs, without acquiescing the role played by a multitude of sociocultural factors such as family, regional background, and caste–class position in deciding 'employability'. This illusion has spurred investment in engineering education not only by individuals, but also by various sociocultural organizations, caste associations, and the postcolonial state in the form of subsidies for disadvantaged groups. Such notions have also been instrumental in producing education as a 'contradictory resource' (Willis 1977) responsible for differential outcomes for students, resulting in crippling immobility for some and monetary gain, prestige, and success for others.

The narrative of education leading to success had its roots in the twentieth century when several elite communities in South India were able to use their accumulated resources from agriculture and small businesses to seal their rural–urban dominance by accessing English education (Baker 1984; Upadhya 1997; Subramanian 2015a; Fuller and Narasimhan 2008a). Upadhya (2016a) describes such changes as the successful capture of 'information goods' by landowning castes, through which they not only reinforced their status but also secured a certain visibility through their accumulation of various kinds of cultural, social, and symbolic capital (also see Upadhya 1997). Influenced by the career trajectories of relatives and other young men from their respective communities who have been able to successfully access white-collar employment through higher education, agrarian and industrialized middle-castes in many parts of the country have also aspired to diversify into professions not affected by the seasonal instability of agriculture and business (Upadhya 1997; Chari 2004a; De Neve 2006, 2011; Jeffrey, Jeffery, and Jeffery 2008; Vijayabaskar and Wyatt 2013). Investment in education, especially in professional courses such as engineering, medicine, and management appeared as a naturalized, if somewhat expensive, strategy of consolidating middle-class success, especially for sons (Upadhya 1997; De Neve 2011). The degree is important not only for securing a livelihood, but also as a form of symbolic capital—to appear 'cultured', command respect in society, and useful for securing substantial dowries in the form of cars, gold, and cash (Upadhya 1997). Though college education for daughters, including in professional courses such as engineering, has become more widespread since the 1990s, it is also driven by the interests of securing the new norm of professional hypergamy, where engineer grooms seek similarly qualified brides.

Individual communities' strategies involve not just investment in education for their children, but also the establishment of institutions and political action for greater representation in affirmative action policies (Upadhya 1997; Wyatt 2010; Vijayabaskar and Wyatt 2013). While gendered patterns of enrolment in education are intended to protect and enhance family respectability, political movements are couched in a language of state neglect and competition against Dalit groups. Middle-castes perceive Dalit mobility as a threat, due to the reservation policies granted to them since independence, as well as anxiety over losing control over their labour (Gorringe 2012; Vijayabaskar and Wyatt 2013; Chowdhry 2007). The animosity against Dalit presence in spaces like engineering colleges also has a gendered angle as evident in slogans such as 'First our jobs, then our girls'—which have gained ground amongst middle castes not only in the south, but also in northern India (Chowdhry 2009). Studies have also shown that men of Dalit and Other Backward Class (OBC) castes are engaged in a vicious cycle of gendered violence on college campuses while caught in a race for white-collar employment and mobility that pits them against each other, even while facing similar disadvantages such as lack of English skills and other forms of cultural capital valued in employment (Rogers 2008).

These dynamics of caste, class, and gender, lie at the heart of private engineering education and animate everyday interactions at the colleges. Just as the gigantic structures that house engineering colleges represent the scale of aspirations, anxieties are also spatialized on campuses. Though most professional colleges are co-educational following the public university system (rather than the gender-specific model established by European and American missionaries in the early twentieth century(see Burton 1996; Krishnan 2017), they are characterized by strict gendered rules for student conduct, adherence to which is secured through a signed legal document during admission. Many of the colleges emphasize modest dressing covering legs, arms, and chest (A. Ram 2015), strict curfews, limited outings, as well as scrutiny of cross-sex interaction; in some colleges, men and women are not allowed to communicate at all. Mobile phones are banned. Such rules are in stark contrast to the lack of regulations in public universities, and private colleges attempt to out-do each other in the conservatism of the codes of conduct imposed on students. These rules and regulations are also meant to quell the insecurities of parents who are afraid of *children misusing their freedom* and view the spaces of higher education as a 'political liability' (Mukhopadhyay 1994; also see S. Krishnan 2015) because of the possibility of young people having sex before marriage, or a 'love marriage' across boundaries of caste.

Teachers and members of the non-teaching staff are tasked with the labour of policing students' interactions with the opposite sex along with the deployment of various technologies of surveillance such as closed-circuit cameras and 'smart' identification cards to be swiped on entry and exit. Many colleges have incorporated features of gender segregation into their architectural design in the form of separate staircases, canteens, and corridors, and even enforce separation on college buses, with chains running along the length of the interior. Students who do not follow these codes or are observed engaging in 'undesirable' behaviour such as being close or intimate with a peer of the opposite sex are called for 'counselling', or even physically punished in some institutions. In other words, parental anxieties regarding altered norms of sociality in higher education have imbricated well with the familial- and caste-driven modes in which institutions are run, strengthening disciplinary apparatuses and mechanisms of surveillance that stress temporal and spatial boundaries for young people, especially women.

These gendered conditions[6] under which young women are allowed to gain access to higher education are important in showing the hollowness of celebratory accounts that suggest that women enter higher education in parity with men (for instance Clark 2016), or even the state-authored discourse that women in the spaces of Information and Communication Technology (ICT) are an essential part of contemporary Indian as well as 'Tamil modernity' (Pal 2019). In fact, the strictures in colleges establish the tenacity of the 'patrifocal structure and ideology' of the Indian family that Mukhopadhyay and Seymour (1994) identified as being the chief obstacle to Indian women's advancement in science and technology careers, in their study conducted in the 1980s. 'Patrifocality' is the tendency in Indian families to prioritize a son's education and career over a daughter's, and the husband's over the wife's, because of which fewer women can sustain their careers. Despite the feminization of labour[7] that industries like

---

[6] This is not specific to India. In the West, where the concern with gender and science is widespread among feminists, it has been observed that girls and women drop out of science, technology, engineering and math (STEM) careers more than boys and men at various levels, right from school. This has been referred to using the metaphor of a 'leaky pipeline' (Blickenstaff 2005) and has been attributed to cultural perceptions that careers in these subjects are 'masculine', unsuitable for women, and negatively impact notions of femininity (Seymour and Hewitt 1997). Investigating the phenomenon further, Jacob Blickenstaff (2005) found that several gender-based 'filters' exist 'in the pipeline' systematically keeping women out. These include pedagogical approaches, sexist attitudes of teaching staff, pressure to fulfil gender-based roles, and the inherent economic value attributed to branches of science such as physics and engineering in a male-dominated society. Women also reported a 'chilly climate' while pursuing careers in science and technology—reporting experiences of 'isolation', 'psychological alienation', 'male (masculine) attitudes in groups', and 'intellectual intimidation' (Brainard and Carlin 2013). Melanie Walker (2001) argues that this is because engineering is a field in which gendered identities are constructed in particular ways: the language of technology is masculine and the discourse of engineering is one of technical competence and expertise expressed in masculine terms. Therefore, despite increased numbers in engineering colleges that challenge men's occupational segregation as engineers, dominant relations are left in place. Carol Mukhopadhyay (2009) finds a comparable context in India, in terms of the number of drop-outs. Yet, one cannot simply import Western theories of gendered science to apply to the Indian situation, she argues (2009, 143). Though connections to the First World exist, the lives and careers of women in science in the Third World reflect 'a curious intermingling' of the many contradictions in these societies (Subrahmanyan 1998, 19). Although women have made significant strides in entering higher education in the past two or three decades (Chanana 2001; Mukhopadhyay and Seymour 1994; Seymour 1995; Subrahmanyan 1998), women's participation in the labour force remains as low as twenty-five per cent (in urban areas, as low as 15.4 per cent) (2011 Census).

[7] The term refers to women's increased participation in paid work as well as to the deterioration of working conditions in previously male jobs.

IT are said to represent, only a highly fragmented section of women from the Indian middle class have been able to successfully transition from higher education (including in courses such as engineering and management) to sustained employment. Indian women's participation in the labour force is amongst the lowest in the world and predicted to decline further (NSO Survey 2019–20). Even as the family structure plays a role in restricting women from pursuing employment opportunities, this book shows that the content and form of higher education is also instrumental in legitimizing barriers and boundaries, limiting the ways women imagine their career trajectories, and the realm of what is possible for them. The crux of higher education for women, as imagined by those in authority at least, seems to be to accumulate credentialed capital without ever acquiring a sense of agency or autonomy. Yet, as I elaborate in the forthcoming chapters, these rules imbricate almost seamlessly with rules meant to make them more 'professional', and women do dream of jobs in tandem with their 'intimate aspirations' for marriage, love, and romance. Young people's negotiations with the forms of control described earlier, their subjectivities, what constitutes their agency, and their experiences of the life-phase, youth within such highly policed settings form a significant part of this book.

However, while grappling with young people's aspirations and hopes, it is important to note that opportunities are not tantamount to the levels of aspiration. The dotcom bust of the early 2000s and outrage against outsourcing in developed countries has cut into opportunities available for engineering graduates: unemployment and underemployment are rampant. The Information Technology industry, the largest recruiter in colleges, has always maintained that despite these engineering graduates being technically strong, most students lack the communication and language skills required to service offshore clients, and are therefore 'unemployable' (see Upadhya and Vasavi 2006).

The sense of disappointment from the employability crisis has somewhat dampened the aspiration for engineering, and a growing percentage of places in engineering colleges have been going empty since the early 2010s, reaching an all-time high of fifty-two per cent in 2019 (Sivapriyan 2019). This is also because the regulatory authorities overestimated the proportions of the IT boom and overissued licences to start new colleges. Following the employability crisis, there is widespread consensus among various authorities that there are about 200 more colleges than needed in Tamil Nadu. To stem the excess, the AICTE had curbed the issue of new licences until 2022, some colleges have been mandated to shut down, and others asked to lower the number of places (Sharma 2021). This downturn also reflects the fleeting nature of privilege in global capitalism, in which certain disciplines and streams are in high demand for a while, only to be rendered obsolete in the process of transition to greater automation. Yet, with BTech degrees now being offered in newer fields such as Business Analytics, Artificial Intelligence, and Machine Learning, which are touted to be the next big thing, engineering colleges continue to be the crucible in which middle-class aspirations for white-collar employment are forged.

Family, Private Education, and the Modern Nation State

In the popular Hindi film *3 Idiots* (Hirani 2009), one of the main characters Farhan Qureshi (actor R. Madhavan) introduces himself saying he is modelled on the quintessential middle-class dream that the son would become an engineer, so much so that '*Mera beta engineer banega*' ('my son will become an engineer') (Hirani 2009, 00:10:57) were the very first words uttered by his father upon seeing him as a newborn. This cinematic moment is a good starting point to understand how parents' roles are deeply implicated in producing engineering as the de facto career choice, the only decision commensurate with the sacrifices they have made; a choice perhaps surpassed only by the decision to become a doctor because of the associated status of these two professions. To secure this goal, parents are also keen to send their children to the most competitive institutions, spend huge sums of money on the 'shadow economy' of coaching classes (Cross 2013; Ørberg 2017), and resort to all kinds of threats, physical punishment, and 'emotional blackmail' (Kumar 2011) to make the child 'succeed'—a term often narrowly interpreted in terms of professional degree, white-collar employment and securing the trappings of middle-class lifestyle. These 'educational dispositions' (Kumar 2011) are important in highlighting the operative praxis of the middle classes (Bourdieu 1990), even though the middle classes in India are a rapidly growing and highly differentiated class (Ray and Baviskar 2011; Sridharan 2008; Fernandes 2006).[8] The pressure to perform in Class Twelve (the qualifying exam for professional colleges) is so high in many middle-class households that David Sancho evocatively calls it 'the year that can make or break you' in his book title (Sancho, 2012)!

Within middle-class families, these pressures are construed as 'love', underlining the importance of bringing an affective model to bear on totalizing models of discipline. They constitute the 'childhood *habitus*', which in the Tamil context is often expressed as an 'incompleteness of self' finding 'fulfilment in kinship' (Trawick 1990, 143–145), playing out into adulthood. Thus, my interlocutors did not speak of 'family pressure' in disciplinary terms, but instead remembered the endless cups of tea made by their mothers so they could be awake all night to study, or the times when their fathers pulled extra shifts to secure money for school fees. These were also the memories and bonds invoked when interlocutors felt compelled to follow college rules, secure good marks, or give up their choice relationships in favour of a marriage alliance favoured by parents. Their place in college did not appear to be just theirs but manifested as a family project put together by parents 'toil' as much as their efforts.

As *3 Idiots* (Hirani, 2009) itself dramatizes quite powerfully through the leitmotif of suicide,[9] pressure from the family continues well into engineering student life, sometimes compounded by the need to ease

---

[8] The debate on who constitutes the middle class is not a question I choose to pursue through this book as it is both nebulous and a hotly contested territory (Donner 2011; Ray and Baviskar 2011; Säävälä 2012). I keep in mind only a broad working definition that comes from Bourdieu (1977): my interlocutors belonged to a class that had a positive disposition to acquiring higher education (*habitus*) and had the financial capability of investing in it (*capital*) and acted upon this in constructive ways.

[9] Suicide appears as a shorthand in the film to convey characters' assessment of their failures, and their inability to accept them: thus, one of the main characters in the film, Raju Rastogi (Sharman Joshi), attempts suicide when he is rusticated from the college for drunken misconduct in the classroom (Hirani 2009 1: 47: 27). He is shattered by the prospect that he will not be able to rescue his family from debilitating poverty (father is paralysed, sister is in need of marrying, mother barely manages to make ends meet with her schoolteacher job) (Hirani 2009, 1: 47: 30). Another character, Joy Lobo, is shown hanging from the ceiling fan in his room as he is unable to figure out the mechanics for a drone prototype in time for the deadline of the final year project (Hirani 2009, 00: 34: 05). Third, in a run-up to the film's denouement, it is revealed by the principal's daughter (Kareena Kapoor) that her elder brother committed suicide because their father pressured him to prepare for the engineering college entrance, rather than pursue literature as he desired (and did not die in an accident, as the father has been made to believe) (Hirani 2: 17: 25).

financial burden and secure class mobility. Moral interpretations by college authorities as to what kind of traits constitute a good student or how a student should approach academic life imbricate closely with parental expectations, producing a somewhat homogenous student culture centred on rote learning and reproduction of prescribed texts, with little to show in terms of creative output, a vibrant social life, or critical engagement with larger social and political issues. Students who do not comply with this institution-led study culture are quickly labelled in deviant terms, subject to disciplinary action, and moral judgement. These discourses that centre on academic excellence as the only objective of entry into college have also been important in drafting a model student subjectivity along technocratic lines. These imaginaries leave little room for individual agency or variation, which are conceptualized within the technocratic ideology as being 'prone to human error', detracting students from achieving the end goal. As is coded into machine logics, the end is to produce sameness with consistency, rather than 'risk' failure. It is also to be noted that they are distinctly organized along a binary (success–failure), following the cultural construction of cyber technology (1–0), rather on the experimental and rational basis of science for which, philosophically speaking, there are hundreds of outcomes.

This technocratic vision of student life carries powerful currency within the Indian middle classes, with an inherent bias of caste and gender. This was powerfully demonstrated in the middle-class backlash against student activism in Hyderabad Central University after Dalit student Rohith Vemula's suicide in 2016,[10] and the continued vilification of student leaders and politics as being *unproductive* or *against development* (also see Lukose 2009 for precedents to this discourse in Kerala). During the height of the student protests in 2016, messages circulating on social media explicitly juxtaposed the trajectories of engineering graduates who are able to secure a job by the age of twenty-one or twentty-two with the

---

[10] Though Vemula's suicide was a much-politicized event in the media, there was largely indifference surrounding Dalit students' suicides at Indian Institute of Technology (IIT) campuses s, which has been a regular occurrence since the early 2000s. This has called for an introspection into 'invisible' workings of caste in spaces like IITs, and the inherent casteism in terms like 'merit' which carry much weight in these spaces. See Subramanian (2015a) and Odile and Ferry (2017) for upper-caste bias in places such as the Indian Institute of Technology—Madras, and the relevance of the local in the propagation of caste bias in state universities in South India in a paper by Malish and Ilavarasan (2016).

trajectories of students pursuing advanced research degrees in liberal arts and social sciences on research fellowships supported by the government (as was the case of Jawaharlal Nehru University (JNU) student leaders such as Kanhaiya Kumar). These tend to see white-collar employment as the end goal of student life, rather than the pursuit of research or engagement with politics as being educational or having social value. In fact, the middle-class distaste for student politics was granted so much visibility during this time that questions such as 'Should Attendance be Made Compulsory in JNU?' were posed during nine pm televised debates as if they were genuinely pressing questions of national concern. These discourses have made public universities such as JNU appear as an undisciplined and sexualized foil against the familial regimes of discipline in private colleges. Thus, JNU has been described as a university with condoms and liquor bottles strewn about (Khan 2016), and Kashmiri student Kamran's suicide in the English and Foreign Languages University, another central public institution, was attributed to mental stress emanating from his alleged homosexuality (Henry 2013).

These responses to political events should also be construed as an extension of the 'educational dispositions' (Kumar 2011) forged in the middle-class family, which translate into biases against politics and people perceived as not coming from similar culture with its inherent caste–class bias. Former Minister for Human Resource Development Smriti Irani's reference to twenty-eight-year-old Vemula as a 'child' in Parliament instantiates this (Sen 2016). Thus, public universities which have a culture of vibrant student politics are simultaneously construed as an elite space of (left) liberal intelligentsia, working on topics[11] beyond the relevance of the average middle-class Indian, but also spaces where 'non-meritocratic others' (such as 'lower castes' and Muslims) vie for power. These discourses tacitly endorse the disciplinary regimes followed in institutions such as the private engineering colleges I describe, as the legitimate model of 'productive' education, over the model of education endorsed within a liberal arts or civic framework.

It signals the constitution of privatized education as a moral community organized around a technocratic vision of development but with

---

[11] A message listing various topics of study among JNU students, particularly themes of sexuality and Mughal rulers also made the rounds on social media.

familial, gendered politics at its core—a model that retains continuities with the idea of a nation state as constituted by gendered public/private (Chatterjee 1989), with education firmly situated in the private with all its moral and gendered import. In contrast to the liberal cultures of public universities, the high walls and barbed wires of private campuses are perceived to prize middle-class moralities, and the grounds within emerge as the legitimate soil on which the seeds of middle-class aspirations can be planted without loss of respectability. This underlines the point that the Indian middle classes are not just concerned with securing prestige through education, but also respectability by laying claim on gendered moralities (Säävälä 2012; Gilbertson 2018).[12]

The growing popularity of these discourses, with its inherent gender-caste biases, show the growing preference among middle class for an education that reiterates technological supremacy of the nation, while retaining a moral and social order that is described as 'appropriately Indian' (Radhakrishnan 2011). Though the primacy given to market needs and reconceptualization of education as shaping a workforce for the New India as a 'knowledge society' (Nisbett 2009) have had a role to play, the legitimacy accorded to technocratic visions of society by the middle classes go back to Nehruvian times.

Satish Deshpande (2003) argues that immediately after independence, terms such as 'nation building' and 'development' were not only important, but carried a strong moral impetus that was meant to craft a unified polity. This vision for the nascent country anointed professional experts such as engineers and scientists as the leaders of a new future (2003, 144). The consequent expansion of higher education, especially technical education, and the strong emphasis on scientific and technical training swelled the ranks of the middle class. Though appearing unmarked, this modern middle-class subject carried the premodern privileges of community, caste, gender, and religion, which education was supposed to convert to a 'rational man' with 'scientific temper' (Srivastava 1996; Deshpande 2003). Though this Nehruvian consensus is now considered to have 'broken' with the

---

[12] Ideas about respectability in the middle class emerged in historically and socially contingent ways to inform appearance, conduct of women, and separate them from morally ambiguous lower-class women (Poovey 1984; Walkerdine 1989; Ware 1992; Skeggs 1997). Distancing women from paid labour has been a particularly potent way of class-making (Vera-Sanso 2006; Donner 2011).

contestations posed by various 'alienated' groups rendering different fissures visible (Subramanian 1999; Menon and Nigam 2007; Ilaiah 1996),[13] it should be noted that the technocratic vision remains resilient even in the hands of the Bharatiya Janata Party (BJP) government bent on erasing Nehru as a national and cultural symbol. This is visible in the cultivation of Prime Minister Narendra Modi's image as a man who is in sync with the tech cultures of these times—an image manipulated to showcase his accessibility in a democratic nation state, as much as to inspire a positive attitude from model citizens to technological innovations. Within such a framework, private universities with their surfeit of technology solutions emerge as the cradles of technical citizenship endorsed for the development of the country.[14] In fact, the prominence given to images and quotations of people such as former President of India Dr A. P. J. Kalam in these spaces[15] represent the continued moral privileging of a masculinist technocrat as visionary leader. It also shows the easy conflation produced between science and technology, governance, and technocracy. Kalam's personal project to make students pledge to 'hoist and fly the national flag in their hearts' while 'acquiring knowledge', 'becoming energy independent', planting trees', 'cleaning my town', and 'not spend my winged time in vain' (Kalam website n.d.) presents a kind of productive citizenship that merges entrepreneurial youth with the nation-body (also see Gooptu 2013).

These discourses have melded seamlessly with the dominant discourses in engineering colleges: the structural mechanics of the engineering sciences lend themselves almost naturally to the metaphors of nation building, just as the imagination of the 'networked society' (Castells 2000) is perfectly captured in the complex mechanics of circuits and cybernetics. Engineering graduates, therefore, appear in sync

---

[13] This has also been accompanied by increased mobilizations by backward-class groups for greater representation in politics, education, and employment (Jaffrelot 2003). These new movements significantly changed the composition of the middle class, though older forms of privilege made themselves visible in the assertion of 'merit', cultural capital, and in intangible 'soft skills' (Upadhya 2008; Deshpande 2013; Subramanian 2015a; 2015b).

[14] The anthropology of education has shown education to be responsible for inculcating 'reigning ideologies' (Hall and Jefferson 2006[1975]; Willis 1977; Srivastava 1996; Thapan 2006).

[15] Kalam's portraits are prominently displayed at several events I attended as part of fieldwork; his words were also quoted often, sometimes even displayed along with the portrait itself. Students also quoted him in classroom exercises, often using adjectives such as 'great' and 'visionary leader' to describe him. His portraits were also visible in arts and sciences colleges that I visited.

with the dominant discourses of the times, even as they emerge with signs of cosmopolitanism such as 'exposure' (Fuller and Narasimhan 2006) and training in good communication skills. A gendered sense of morality, which private engineering colleges have further evolved to provide, lends further impetus in securing respectability. The 'moralistic orientation' of technical schools is also meant to build on traditional views of careers in science and technology: engineers are seen as 'men of character' (Ramnath 2017) and a science/engineering degree symbolizes not just general academic achievement and intellectual capacity, but also 'desirable family and personal characteristics' such as 'perseverance, responsibility, filial piety, respect for learning, and ambitiousness' (Mukhopadhyay 1994).

In other words, I argue that the kinds of technologies deployed, and the immense social value placed on technical education as a viable route to a good life, a respectable middle-class identity, and high status have contributed to a specific kind of subject-making on engineering campuses across Tamil Nadu. While enforcing a performativity of the acceptable in the social realm, the institutional structure also propagates a pedagogy that is uncritical and mechanically derived; a pedagogy that emphasizes application over criticism, and structure over individual. Though students are not simply 'interpellated' by this pedagogy and disciplinary framework in an Althusserian manner (Althusser 1971), it does have a role to play in crafting a particular kind of subjectivity and sense of self-worth. Even as students negotiate these institutional constraints, it has become an integral part of college life and the ways in which youth is experienced.

In fact, I would even argue that private engineering colleges have emerged as new important symbols in the weave of the contemporary national fabric, combining Nehruvian 'display of science' (Roy 2007) through futuristic architecture such as large geodesic domes and solar parks with symbols of cultural nationalism and religiosity (neo-Dravidian features and temples on campus). The grid that is meant to shape the new model (male) citizen is no longer confined to 'secular' combinations of modernity, rational thought, and scientific temper as envisioned by Nehru (Srivastava 1996, 2006; Roy 2007), but reworked into a spatial and conceptual idiom of technology intermeshing with non-secular cultures locally.

Though the invocation of structures as shaping the individual within draws from a decidedly Foucauldian imagery of governmentality in which power/knowledge bring about the desired conduct by securing visibility (Foucault 1977),[16] I underline the ways in which the innovations described above consolidate an 'economy of affect', building on Analiese Richard and Daromir Rudnyckyj's (2009) call to examine how 'feelings structure change' (2009, 60). As they posit, affect is crucial in mediating the changed relationship between self, society, and state wrought by neoliberalism by creating an experience of intersubjectivity that 'causes people to create global shifts through their cultural labour' (2009, 60).[17]

In the context of private engineering colleges, considering how an 'economy of affect' circulates to create intersubjectivity is important to understand how transition to privatization has been sustained. An 'economy of affect' (Richard and Rudnyckyj 2009) is also crucial to forging student (and parent) subjectivities that shape the experience of being in spaces of private engineering colleges as something sacred and awe-inspiring; these are institutions run by 'big men' (*periya aal*), to which 'children' (*pasanga*) go to imbibe knowledge, and from which they emerge with status. As subsequent chapters show, the nomenclature, the architecture, and the cultures of patronage in private colleges are all critical in producing these subjectivities. Different colleges can produce variations in affect, depending upon their own histories and the contexts they have created, but ultimately participate in a larger circuit of the affect produced by the aspiration for engineering.

In Tamil Nadu, the affects produced by private colleges mesh with the state's vision for a 'Tamil technocracy' (Pal 2019). As Joyojeet Pal (2019) elaborates, the Dravidian parties, though committed to the idea

---

[16] A Foucauldian framework, where 'knowledge' and 'power' implicate each other, provides a productive lens of thinking through surveillance in the college, and the processes through which young people are measured against certain standards of behaviour that are considered 'healthy', 'acceptable', or 'normal'. However, rather than emphasize the subject as a 'docile body' (Foucault 1977)—an anthropologically unsatisfying premise—I follow Foucault's later work (1988,1990) in which he pays greater attention to the ways in which subjects resist power—focusing on dynamic self-creation and the ways in which subjectivities expand (Foucault 1990, 239).

[17] I find such a methodology of combining discourse and affect useful because it provides an anthropologically productive lens to understand how subjectivities are forged through a series of techniques, processes, and practical action rather than their straightforward 'production' through sweeping economic change.

of the 'organic culture' of Tamil, are also committed to knowledge and modernity. Earth, knowledge, and technology form the tripartite political ideology of the Dravidian parties and are highly visible in government advertisements and other forms of propaganda (Pal 2019, 290). Accordingly, a great number of resources are spent not only in making provisions for IT companies to invest more in the state and providing funds to colleges for employability training, but also for projects such as updating the Tamil lexicon to include technical terms related to computing and launching software in Tamil in order to stay 'relevant for youth' (Muruganandham 2018).

In contrast to the decidedly masculine imaginations of engineering and nation building circulated in the early years of independence, young women are given a special position in contemporary imaginaries: they are part of the marvellous 'modern' make up of buildings that house computers and other forms of technology, but in ways that seem unthreatening to an existing social and moral order (dressed in salwar kameez, dupatta in place). Joyojeet Pal (2019) reads the young women in these state-authored videos as the *mythic* Saraswati (goddess of learning) meant to enhance the appeal of these spaces.[18] Many Tamil films have also followed suit in showing women in software work as being enmeshed in an unproblematic dynamic with men, being able to achieve success while maintaining the desired feminine deportment, and straddling tradition and modernity with ease (Pal 2019; Vera-Senso 2006).[19] This increased visibility accorded to these imaginaries of the 'new Indian woman' (Thapan 2004) are important in fanning aspirations as well as enhancing anxieties related to young unmarried women's presence in spaces of higher education, especially in the face of increased mobilization by disadvantaged groups for greater representation in education, employment, and politics (Gorringe 2012, 2018; Hebbar 2017).

---

[18] The goddess trope has been an important lens for feminist scholars looking at female iconography in media. See C. S. Lakshmi (2008) for an analysis of goddess typologies in Tamil films, and Patricia Uberoi (2002) for a pictorial essay on the ideologies of the postcolonial state in calendar art.

[19] Though this is true in some films, Hebbar (2020) examines certain other films to argue that there is more to the gender politics of aspiration and employment.

Understanding Gender in Intersection with Class and Gender

Complicating gender issues through the lens of caste and class has emerged as a key instrument of analysis to understand contemporary India. Generally known as the intersectional method (Crenshaw 1989), this analysis has been important in showing the interlocking of structures of oppression, discrimination, and politicization in various contexts. Whether this category is useful for the Indian context has been the bone of some contention (Menon 2015; John 2015), yet no one can deny that social life in India presents a complex meshing of privilege and marginality, ritual and routine, official classification, and daily discrimination. Categories such as gender, class, caste, and community cannot be viewed in isolation but as interlacing into a complex grid. In the college context, among other things, it determines who is perceived as gaining legitimate entry into college, their sociality in everyday life, and the limits to their sexuality. These are coded into the concrete structures of the college buildings as much as in the minutiae of everyday social life. They shape discourses and the circulation of affects, distinguishing and differentiating bodies, and determining the conditions of their existence. I feel that an approach that implicitly records these observations as much as individual narratives can rein in the importance given to panopticon-like images of modern institutions or dominant modes of understanding gender and caste relationships as something hyper-visible, spectacular, and incendiary. Instead, the focus shifts to contending with underlying desires, tensions, anxieties, and processes of construction of self within these structures (Kleinman and Fitz-Henry 2007), concentrating on everyday continuities and routines which shape long-standing dispositions as much as epistemic shifts.

The anxieties related to young women in education and employment are not particularly new in Tamil Nadu; some districts, including Salem in which I conducted my research, are part of the 'female infanticide belt' in the 1980s, running from the northwest to the deep south (Chunkath and Athreya 1997). The reasons include reliance on agrarian economy, poverty, financial burdens related to marriage and dowry, gendered patterns of inheritance, sustenance of family name, and the 'social danger' presented by unfettered young women who might 'ruin their reputations' and further burden their families. The last is described by Sneha Krishnan in her study of young women in the city of Chennai, as the anxiety over the 'dangerous instability' (2014, 1) that characterizes female youth. The sense that daughters are a 'liability'[20] is very strong, though there is less incidence of female infanticide and foeticide today.[21] However, as the norm, young women in Tamil Nadu are subject to surveillance throughout their adolescent years until marriage, and daughters are neither allowed to move freely in public spaces nor seek employment unless there is pressing need in the form of extenuating family or financial circumstances.

Similarly, protectionist paradigms are not only enforced in private colleges; they are embedded in the everyday workings of the law, which tends to obfuscate differences between rape and consensual sex, and places women in shelter homes rather than let them contract relationships of their choice when it goes against the family's desires (Mody 2008; Baxi 2014). Similarly, the law as well as various social actors participate in the production of discourses that classify men from certain communities as 'predatory', against whom violence is a pedagogic tool (Rao 2009; Gupta 2001, 2009). The manifestation of these two sentiments is particularly visible in Tamil Nadu since the early 2010s, and violence against inter-caste marriage and associated *'aanava kolai'* ('insolence murders') in the region have sent shock waves across the nation. This violence has particularly targeted young male Dalit engineering college students, their families, and *cheris* (settlements) and are meant to send out a message to Dalits not to breach caste boundaries. Leaders linked to middle-caste parties such as S. Ramadoss of the Pattali Makkal Katchi (a Vanniyar-based electoral party), and S. Yuvaraj of the Kongu Vellala Goundergal Peravai (a Gounder-based group), publicly alleged that inter-caste marriages, especially when Dalit men are involved, are born out of 'caste design' and not for love. 'They wear jeans, T-shirts, and fancy sunglasses to lure girls from other communities', said S. Ramadoss, at a

---

[20] Some alarming practices related to female infanticide and foeticide continue to exist. See report by Menon and Saravanan (2011).
[21] This is generally attributed to former Chief Minister J. Jayalalithaa's Cradle Scheme, through which people were asked to leave their girl children in shelters inaugurated for that purpose rather than end their lives. The scheme was, however, also criticized for doing nothing to alter the mentality that daughters are a burden.

meeting of intermediate castes, which he called the 'non-Dalit block' in 2012 (Daniel 2012).

Caste as sartorial control is not a new phenomenon and has been used for centuries as a method of reinforcing dominance and control (Tarlo 1996; Hardgrave 1969). By attacking emerging sartorial practices of Dalits and their engagement with global youth styles that are not marked by caste, Ramadoss sought to draw attention to their new mobility and class status—enabled by disengagement from regimes of caste-based agricultural and manual labour. This also indicates the slippages between caste and class categories, which are sometimes used interchangeably, because of a strong correlation (even if they do not always hold) (Deshpande 2013; Gilbertson 2018). The statement also highlights the economy of affect surrounding explicitly caste-marked and gendered bodies, the derision felt for what are deemed as *anomalous assemblages* when 'repugnant bodies' are juxtaposed against desirable material goods and commodities. The resulting dissonance has been extremely useful in driving follow-up events, such as getting students to pledge that they would not marry outside their caste (*The Times of India*, 13 October 2012) and politicization of hypergamous marriages of Dalit men (especially young engineering graduates/students) with upper-caste women. The most prominent of these cases was the marriage of Divya, a Vanniyar woman, with a Dalit, Ilavarasan, following which Divya's father committed suicide and three Dalit villages in Dharmapuri district were burnt as retaliation. Following these incidents, Divya chose to return home to her mother and Ilavarasan was found dead by the railway tracks. Though many Dalit activists alleged murder, other theories circulated too, such as what Ramadoss alleged about Dalit men 'trapping' OBC women.

In characterizing such episodes as mere strategies of 'political entrepreneurs' (Wyatt 2010) to earn political mileage, we lose an important aspect of how caste circulates as an 'affect' in encounters with other bodies, or how certain views become 'shared perceptions' (Ahmed 2004). That caste is an important marker of drawing affect, particularly through the trope of sexualized bodies and anomalous assemblage has also been made explicit in a feature film called *Draupathi* (Kshatriyan, 2020), which means to 'reveal' what Ramadoss had alleged: the film showed gangs of Dalit young men being equipped with iPhones and motorcycles by their leaders, so they can woo OBC women, marry them in secret, and 'use them' to harass OBC families. The film also showed the breakdown of caste-based regimes of labour and attributed it to government schemes such as MGNREGA[22] (The Mahatma Gandhi National Rural Employment Guarantee Act) portraying them as 'easy money' granted to the poor irrespective of whether they work or not. According to the film, this is the money that funds 'scams' like the harassment of OBC women. Entirely crowd-funded with a largely unknown cast, the film has been the sleeper hit of 2020, with movie halls running to packed audiences.

'Caste is the feeling that overwhelms you when you see your community's beauty, its oneness, its culture, or when you feel bad that it is all disappearing with modern times and Western culture', one of my interlocutors summarized when discussing the Divya–Ilavarsan case. 'It is bad perhaps, but it is there', Uthra, another interlocutor said. Responses such as these are tacit acknowledgement that caste, though delegitimized in the public domain, is still an important part of the interiority in the ways it is practised as tradition at home, as loyalty to a certain party with its culmination in a vote bank, or as the ability to read cosmopolitan markers as the basis for caste identities.[23] Such assertions of caste identities are not isolated, but visible across the country. Gendered slogans are stamped on vehicles. Agitation by various 'dominant castes' often disrupt highway traffic in demonstrations of 'hurt pride'. These communities, though doing well on various indicators such as income and employment, many on par with the Brahmin communities in their respective regions, have been fighting for inclusion in the lists of OBC (Other Backward Class) to access reservations in higher education and public sector employment, based on the feeling that they are disadvantaged (Deshpande and Ramachandran 2017).

---

[22] The Mahatma Gandhi National Rural Employment Guarantee Act, 2005, is an Indian labour law and social security measure that aims to guarantee 'right to work' by providing at least 100 days of wage employment in a financial year to every household.
[23] In Telengana and Andhra Pradesh, a Telugu song called 'Caste Feeling' in *Amma Rajyam Lo Kadapa Biddalu* (2019) directed by Sidartha Thatolu made waves on social media. Sung by Ram Gopal Varma in a staccato voice that is more political chant than musical, 'Caste Feeling' is meant to be a vindication of casteism. Roughly translated, it says that we all think of 'caste feeling' as something to be kept locked inside because the laws of the land ask us to do so, but the Constitution itself enshrines caste, which is why there is nothing wrong in wearing your caste identity in public. The song is accompanied by visuals of various politicians from the Andhra Assembly with their caste names duly highlighted, ostensibly to show their vote banks. See https://www.youtube.com/watch?v=JykSPfmaZRc.

These incidents show the continued resilience of caste, despite the modern nation state's aspirations to do away with caste hierarchy as stated in Article 15 of the Indian Constitution. Sociologist Surinder Jodhka argues that it has persisted as a 'system that institutionalizes humiliation as a social and cultural practice' (Jodhka 2015a, 12), even as it has changed its 'avatar' as a closed system of ascribed status (Jodhka 2015b; Abraham 2014). Thus, even if 'boundaries' are more porous than before, and it is possible for achieved status to trump ascribed statuses, 'caste' continues to be more visible even if less relevant (Jodhka 2015b). These factors have led to its conceptualization as a form of 'capital' which consistently shapes work and employment opportunities for its members (Gilbertson 2018, 116).[24] It has also become well entrenched in the private sector, despite the latter's claims to being 'caste blind' and 'meritocratic' (Jodhka 2015a; Jodhka and Newman 2007), as well as in politics, despite being formally delegitimized according to the law.

Following these, many have theorized that a non-hierarchical and egalitarian society does not quite exist in the Indian world view (Gilbertson 2018, 201), which is why caste as a system of continued inequities continues to mutate, working itself into secular tropes, becoming encoded in everyday language and behaviours. This can be seen in the ways that 'merit' has come to stand for 'upper caste' just as 'reservation' has come to stand in for 'lower caste' (Deshpande 2013; Subramanian 2015), or the preference for 'professional hypergamy' (Baas 2009, Kalpagam 2008), which couches caste preference in the language of similarity of work cultures and class. Similarly, political action for caste mobility is often articulated as 'regional aspiration' (Vijayabaskar and Wyatt 2013), or 'cosmopolitanism' in the cities is articulated as 'castelessness' though often refers to the upper castes (Gilbertson 2018). These reflect the reworking of caste into a secular idiom of 'culture', reflecting the French sociologist Louis Dumont's arguments on the transformation of caste from *jati* (a ritual category with links to occupations in a primarily agrarian social order) to a secular entity as ethnicity, culture, or *samaj*

---

[24] In contemporary India, caste is an avenue to access and allocate resources, is embedded in politics and law (Mody 2008; Baxi 2014), in contentions around the definition and categorization of communities (Washbrook 1975, Dirks 1992), and in spaces of education and employment (Subramanian 2015a; Malish and Ilavarsan 2016; Jodhka and Newman 2007.

(society) with applications in modern and urban contexts (Dumont 1980, 221; Mines 1996; Natrajan 2012).

Yet, this does not explain the intense 'affect' that caste has on its members, in forging a collective identity—the capacity to incite violence or to inspire pride and love, to find kinship amongst members, fund projects for their 'development', nurse aspirations for the collective, or feel threatened when faced with competition, and so on. To contend with these multifarious ways in which caste works, one needs to place at its heart the power of 'discourse-affect' in bringing about 'caste feeling'. Though 'caste feeling' understood as just a 'feeling' or 'emotion' conveys something 'personal' and 'subjective', it is important to underline that the social and political are deeply implicated in the term 'caste feeling', which is why 'affect' is probably more accurate than just 'feeling'.

'Caste feeling' not only encapsulates the contemporary backlash against Dalits for state benefits granted to them, but also lends itself at a conceptual level ('caste as feeling') to accurately sum up the complexities of caste in India today, especially as it relates to higher education. This is not to say that there is no discrimination based on caste in higher education, despite the perception that higher education can level chances of mobility between those who have inherited power and traditionally disadvantaged groups. With the implementation of reservation for OBCs recommended by the Mandal Commission in the early 1990s, public-funded institutions of education have regularly transformed into theatres of protest—both in favour and against the reforms. Those in favour of affirmative action policies for Dalits, Adivasi, and other traditionally disadvantaged groups have often spoken of their experience of humiliation to bring caste issues in higher education to the fore. For subjective accounts of caste and higher education, one must look at accounts such as N. Sukumar's (2013), on his life as a student in Hyderabad Central University, and the furore around the death of Vemula in February 2016 at the same university. A rich literature on experiences of caste in higher education highlights many complex issues around caste in higher education, right from the time of admission to the everyday life to the specific forms that education takes when caste is involved, and the everyday marginalization that students from these communities face (Deshpande and Zacharias 2013; Vasavi 2006; Tharu 1998).

Within this body of scholarship, caste and technical education specifically is also now a distinct branch of scholarship within the social sciences. Specific studies have attempted to look at the lived experience of caste in engineering colleges, following what has been seen as an 'epidemic' of Dalit suicides in elite institutions such as the IITs as a result of casteist assault. Ajantha Subramanian (2015a) finds that elite institutions such as the IIT Madras have become a means of 'recovering caste privilege' for privileged castes such as the Brahmins who use discourses such as 'merit' to assert their caste privilege in 'casteless' terms. Following the framework laid by Satish Deshpande (2013), Subramanian draws a startling picture of the way caste hegemony is reproduced in such institutions.

However, Malish and Ilavarasan (2016), in the Kerala context, find that institutional cultures in centre-run universities have a better *institutional habitus* for Scheduled Caste students than state-run colleges; the former provide a more conducive environment for students of different educational and familial background to take up higher studies, because the primary division between students is those who are from the state and those who are not. Thus aspects such as caste are not as well recognized, whereas in the state-run institutions, which have more local students, aspects such as caste identity take the forefront.

Despite a few encouraging signs, studies show that higher education has not really been able to secure mobility for Scheduled Caste and Scheduled Tribe students. A recent study by Odile and Ferry (2017) in IIT Lucknow highlights 'the processes by which the IITs both continually and differentially eliminate students from the dominated groups (particularly those belonging to the SC and ST categories), and contribute to a strongly differentiated return, hence the social value, of academic titles on the job market' (2017, 2). Their study conclusively shows that contrary to popular perceptions, it is students from reserved categories who suffer unequal treatment for equal marks. Market logics dictate that it is invariably those with 'upper caste' markers such as success in extra-curricular activities and good English-speaking skills who are recruited by the private sector. This has been conclusively shown by studies of recruitment practices in education and private sector (Jodhka and Newman 2007), where caste capital often translates into modern capital.

Despite these insightful studies on marginalized groups in engineering institutions, the 'rise of the backward castes' remains under studied in the higher education context. Though studies have shown that unemployment is high among educated Jats in Uttar Pradesh (Jeffrey 2010; Jeffrey, Jeffery, and Jeffery 2008), and the caste intersects in a very complex way with globalization (Lukose 2009), there has been no in-depth study yet of the nature of the proliferation of private engineering colleges in South India and the role played by middle castes, caste associations, and 'big men' in crafting the upward mobility of communities through private higher education.

My objective is to capture the intense feeling that derives as much from a sense of superiority, pride, or tradition as the sense of being locked in a collective and intense competition with various caste groups, producing a deep affect that drives fights, quarrels, and rivalries and marking of gendered boundaries. In conceptualizing caste as feeling, I follow affect theorists who understand affect as 'energies transmitted through bodily encounters' (Åhäll 2018, 40), in ways that reinforce prejudice, or reaffirm pride, so that they become 'judgements that hold' or 'become agreed as shared perceptions' (Ahmed 2014, 208). It enables a shift from thinking of it just as an individual feeling to implicate the collective.

Within such an understanding of caste, youth are very much seen as part of caste society, even as its agents: to be bequeathed with education through trust funds; whose entry into educational institutions is to be lobbied for; who are in need of 'guidance' and 'counselling' to make the 'right' choices in sociality and in marriage; to be schooled into a profession with all the desirable traits of a middle-class identity so the community can achieve mobility. Youth, too, propagate these ideas through their practice of status and prestige-building practices within their social groups and the enforcement of moral boundaries as a form of 'peer pressure' (see Chapter 4 in this volume).

Mapping gender, class, and caste in private engineering colleges, therefore, shows the different ways of being 'youth' in South Asia, despite attaining the legal majority at eighteen years of age. The debates contained in the book intersect with discussions of youth in cultures of study (Sancho 2013; Ørberg 2017; Odile and Ferry 2017), 'college culture' (Osella and Osella 1998; Nakassis 2016), youth in politics (Gorringe 2012; Jeffrey 2010), as engaging in status-diminishing practices (Nisbett

2007; Nakassis 2016), transgressions of boundaries (Heitmeyer 2016; Chowdhry 2007), or being involved in the search for a spouse or romance (Lukose 2009; Clarke-Decés 2014; Bhandari 2017).

In the higher education arena, I identify four main tangents along which these intersections have developed, and which I intend to examine in my chapters:

- Processes of staking ownership over education through private enterprise and mobilization of various kinds of capital
- Facilitating access and drawing limits to enrolments and opportunities in education in terms of class, caste, and gender
- Deciding employability
- Determining 'suitabilities' and 'subjectivities' in the construction of heterosexual love, romance, and possibility of marriage.

Yet none of these is fixed or set in stone. Despite the narrow and conservative connotations of these terms, each chapter captures a sense of flux, of constant change, and shifting boundaries. In the first two chapters, we see that due to multiple interrelationships between caste and business networks as well as dependence on inter-caste formations of labour 'Institutional' big men realize that a focus purely on the advancement of a particular caste group is 'no longer feasible or desirable' (De Neve 2000, 516). Moreover, institutions of higher education are not allowed caste-specific admission. Caste-specific patronage has therefore been worked into an expansive discourse of 'regional development', just as philanthropy has been worked into an idiom of welfare for the whole community. However, the kinds of communities built are centred on various local axes around which giving, taking, and morality make sense. This is made explicit in the latter three chapters while gendered moralities, marriage preferences, and the relevance of local identities in everyday life in the college.

Sites and Methods

I conducted fieldwork in Salem City, a medium-sized town in Salem district, western Tamil Nadu in 2014 at the peak of the engineering college boom. The choice of location was deliberately non-metropolitan to enable a point of view different from the metropolitan cities such as Bangalore and Chennai, which have several economic zones and corridors dedicated to IT, and are known to be transformed by IT culture (Nisbett 2009; Venkatraman 2017). Salem town, located approximately halfway between these two IT hubs, has been slower to catch up. Though influenced by IT culture, the history of Salem is also defined by older industries that had been set up under the aegis of the state (such as the Salem Steel Factory) as well as those that grew organically to process and market products of the region such as cotton and tapioca. Given this backdrop, where homes are built amid areca *thottams* (gardens) and power looms occupy rooms within the household itself, Kongunadu can be characterized as 'rurban', in which farmlands, industry, centres of trade, export, and residences coexist spatially. Salem is also intricately connected to other urban centres in the western districts such as Coimbatore, Erode, and Tiruppur, not only by interests of industry and trade, but also kinship, affinity, and marriage, which is why I lay great emphasis on Kongunadu in the study and not just the city of Salem.

The ethnographic research was conducted over ten months, from February to December 2014, in a self-financed engineering college in Salem district in northwestern Tamil Nadu. I stayed in the college during this period, attended class with the students, ate in the mess, slept in their rooms, and participated in every aspect of their daily lives from attended birthday parties, stealing into the terrace at night for some fun, cooking on the sly in the hostel room, and even going on meetings with boyfriends. Since then, I have kept in touch with my interlocutors through social media and regular phone conversations. I also visited other colleges in the state, including some in Chennai and Coimbatore, and carried out in-depth interviews with members of various college managements. I have also engaged with various material on colleges available in the archive.

The college in which I stayed, which I call Chinna College of Technology (henceforth CCT)[25] for the purpose of this ethnography, is an autonomous institution affiliated to Anna University and situated in

---

[25] Name changed to protect identities.

Salem City. During a reconnaissance trip conducted in December 2013, I visited several colleges in the area to explore the possibility of carrying out fieldwork. Among the colleges visited, the principal of CCT was most welcoming of my anthropological interest in the everyday life of the college and after examining my research proposal (a one-pager that I had written for this purpose) said he had no objection to the research, and invited me to stay in the hostel in exchange for a nominal fee (₹3,000 or $36 US a month) that included meals. I stayed in the women's hostel named after the founder's wife, attending classes, and shadowing my interlocutors (who were between eighteen and twenty-one years old) as a participant observer.[26] I followed up on what I observed through open-ended interviews with my student interlocutors in informal settings, whereas I conducted recorded interviews of teachers and members of the management in their offices. The students at the college (totaling up to about 4000) came from various religious backgrounds, castes, class, and ethnic backgrounds, though the bulk (about sixty per cent) were from Backward Class (BC) groups in Tamil Nadu, in line with reservation policies in the state. The student body also comprised young men and women from neighbouring states such as Kerala, Andhra Pradesh, and Karnataka, other South Asian countries such as Nepal and Bhutan, and African countries such as Somalia and Sudan. These students are lured to Tamil Nadu for an engineering degree from the reputed Anna University, to which CCT is affiliated, because they can access the Non-Resident Indian Quota (NRI Quota) (five per cent of total places). Although engineering education is expensive and colleges are known to charge high 'management fees', students of Scheduled Castes, Scheduled Tribes, and those who aspire to be the very first graduates of their families can avail themselves of subsidies, including free education in private engineering colleges as part of state schemes. Students, therefore, come from various strata of society ranging from those who can pay the high management fees and those who would not have been able to access engineering education without government schemes and bank loans.

---

[26] The ratio of young men was at fifty-seven per cent, marginally more than the number of women.

The Chinna College of Technology

The Chinna College of Technology was founded in the late 1990s by a prominent industrialist from the Nattukottai Chettiar community.[27] 'Chinna' was an endearment used by his nearest and dearest. Already running a successful business in textile manufacturing, he invested in a polytechnic institution primarily to train the labour needed for his factory. As demand for formal engineering training outgrew vocational training given in the polytechnic, an engineering college was set up on the same centrally located premises by his successors.

While the State Highway ran in front of the CCT campus, a narrower road, perpendicular to the main road, bisected the large campus in two—the academic and the hostel sections. The academic section on the right can be entered from the main road through a tall, cast-iron gate, and a wide tarred road leads straight to the Main Administrative Building. On either side, there were well-manicured lawns, frangipani trees, security checkpoints, and marked quadrangles for sports. There were also small notice boards everywhere sticking out of the ground asking students to stay off the lawn. Smaller roads and paths led to various buildings that house classrooms, laboratories, workshops, and offices for the faculty and the non-teaching staff.

Across the narrower road, in the hostel section, I shared a room with a 'tutor-in-charge', a member of the teaching faculty, who lived in the hostel. Chethana is a few months younger than me, had done her BTech at CCT, went to Chennai for a two-year MTech programme, and

---

[27] The Nattukottai Chettiars (Chettiars of the Country Houses) or the Nakarattas have received anthropological interest primarily because of their renown as the prominent banking community in South India, who blossomed under the colonial masters. David Rudner (1994) says: 'although it is possible to trace many Nakarattar commercial practices back to the Chola period, the caste itself does not appear in the historical record until the seventeenth century when they were primarily involved in small-scale itinerant salt-trading activities in the interior regions of Tamil-speaking South India. By the eighteenth century, some individuals had extended their business operations as far south as the pearl, rice, cloth, and arrack trade of Ceylon; others, as far north as the rice and wheat trade of Calcutta. By the nineteenth and early twentieth centuries, Nakarattars were the major sources of finance for myriad agrarian transactions between foreign British rulers and local populations by monopolizing important components of the credit, banking, and agrarian systems of Southeast Asia, and by remitting huge amounts of capital back from Southeast Asia to their South Indian homeland for industrial investment and large-scale philanthropy.' With the growth of nationalist movements, 'elite Nakarattars began a gradual transfer and "freezing" of investment capital from mercantile to industrial ventures' (Rudner 1994, 3–4).

returned as a lecturer. Our room was situated right opposite the office of the chief warden. The room we shared was modest—two cots against two walls, two *almirahs* (wardrobes), and two desks occupying diagonal corners. Next to one of the desks was a large window overlooking an informal quadrangle (not explicitly marked) where the women residents played badminton, preferring this shielded space rather than the large, open grounds and marked courts outside which were dominated by male students. Small ornamental trees were planted along the hostel building; outside our room is a *parijat* tree (night-flowering jasmine) that sprays the ground below with white-and-saffron blossoms and sent a gentle fragrance wafting through the vertical bars of our window. I often woke up to the strains of Saivite bhajans from a temple, with which the hostel shared a wall.

My age at that time, my unmarried status, and appearance helped me merge with the student body, although almost everyone I met placed me as an NRI (Non Resident Indian) student. I dressed in salwar kameez throughout the course of my fieldwork, taking care to wear *bindis, dupattas*, and tie my short hair at the back. However, despite these attempts to fit in, it was obvious to everyone that I was not from there, at least in the initial months. To convince students that I shared their backgrounds, I spoke often of my undergraduate years in Chennai and spoke in the youth slang that I had picked up during my undergraduate years. While the youthful terms helped me find acceptance, I learned that *Madras bashai* was considered somewhat rude in Kongunadu, where the Tamil was almost musical, respectful, and pleasing on the ears compared to the cacophonous lingua franca of the capital. However, I was in for a pleasant surprise as many interlocutors seemed to know and speak Kannada, my native tongue. Salem, as a part of the erstwhile kingdom of Mysore, had many Kannada speakers, and I was thrilled that I could carry out many long conversations in my native language. However, my periodic lapses into English when speaking also helped me make inroads as I was soon sought out for informal English lessons. These informal sessions, carried out in students' rooms after dinner, gave me a legitimate excuse to hang out with students in their private spaces every day and converse with them over a range of topics. In addition to structured interviews with management, fieldwork therefore chiefly involved hanging out with student interlocutors as they went about their everyday lives, attempting to understand their chatter, aspirations, disappointments, plans, and actions. Doing this gave me an overall picture of life in engineering college, its complex nature, fraught with confusion and contradiction as much as conviction and hope. This book aims to capture and analyse what I observed and saw, in all its inherent hues and paradoxes.

In sharing aspects of my own life with the students, I had to share details about my life. A premise of friendship was the primary mode through which I got to know my interlocutors; they were aware that I was writing down detailed notes about their lives and our interactions and made pointed but humorous remarks about it from time to time. For instance, they spoke in laudatory terms that they were the likely protagonists of a book that I would write about them. Despite the authority I now claim to represent their lives, I must also highlight how they represented me from time to time to their friends, classmates, parents, and teachers. In the initial phase of my fieldwork, rumours floated that I was the 'principal's spy'—a researcher hired by the principal to pose as a part of the student body to learn details of students' private lives. Once they got to know me better, students represented me to their parents as an *akka* (elder sister), who would even double up as a chaperone during outings. I was also sought out for careers advice, especially by students aspiring to enter the civil services, or higher education.

Categories in the Field

Given the sensitive nature of my topic and the processes of regimentation I describe, which seem outrageous and regressive within the framework of liberal universities and cosmopolitan locations that many members of the audience at conferences and readers inhabit, I am often asked why there is no similar outrage on the 'field'—'why are young people and their parents tolerating this?'. These questions have been instrumental, after all, in contemporary movements such as Pinjra Tod (*Break the Cage*),[28]

---

[28] Pinjra Tod is a Delhi-based collective that seeks to make regulations for hostel and paying guest accommodation less restrictive for women students. Challenging what they consider to be the CCTV-driven police-security complex, Pinjra Tod demands that notions of safety and security not be used to silence women's rights to mobility and liberation. The movement has garnered momentum across several elite college across the country such as the IITs, the NITs, and Punjab University.

which has spurred many young women, schooled in feminist and gender studies discourses, to speak out against hostel regulations. As a researcher framed within a similar context, much of my time has also been spent in understanding the same—prompting me to dwell and theorize on subjectivities within disciplinary frameworks, and how people make sense of such rules and regimentation. While these discussions are contained in the chapters, a short discussion on how these issues are understood in the 'field' is also called for, to satisfy the basic questions. I will answer these indirectly by describing how I was introduced to these issues in the field myself; such a mode will also illuminate the researcher's subject position within the spaces and discourses of the field. I have also used this as a strategy in the forthcoming chapters to punctuate the authorial voice, with excerpts from field notes.

Given that the topic of my research related to gender and sexuality, the principal of CCT was very keen that I work with Dr X who oversaw the Women's Empowerment Cell, the only official gender-related body within the college. Both during the pilot study and the first few weeks, I was asked to speak to her about 'women's issues' in the college—reflecting the ghettoization of gender into 'women's issues' in much of provincial India. When I opened our first interview with a general question about gender issues in the college, Dr X was excited to tell me that CCT was one of the few colleges in the area that did not explicitly forbid cross-sex interaction and had an environment in which male and female students could talk to one another. She was keen to highlight that a small minority of the student body was not comfortable doing that, but the 'free atmosphere' in the college guaranteed that students would eventually have some form of cross-sex interaction in the college—a 'life-skill' that would also be useful for them in the workplace.[29]

The Women's Empowerment Cell, it turned out, did not have much of a role apart from organizing the Women's Day celebrations every year. It did not explicitly deal with sexual harassment or gender sensitization, and according to Dr X such concerns were not really a priority as they had not had a single incident of sexual harassment in the history of the

---

[29] This remark must be understood in context of the fact that several engineering colleges in Tamil Nadu explicitly forbid cross-sex interaction giving the impression that CCT had a very liberal atmosphere.

college. In fact, Dr X said that it was more likely 'that the girls here will chase boys!'. When I probed further, asking whether they had not encountered any problem relating to gender, she narrated a couple of what she called 'problematic cases' such as that of a young man who was found photographing a young teacher without her knowledge and consent, and a case in which a young man was found to be 'very aggressive' and 'psychic' (sic). These incidents had taken place a few years ago, she said. In the case of the photographs, it had been ascertained that the photograph was 'innocent', whereas counselling was advised for the young man who had demonstrated aggressiveness. 'Other than such issues, we have never had any problems. Students here are particularly good', she said, a term which I understand to be somewhat synonymous with docile and compliant.

Despite Dr X's claims that there were no issues, the above conversation highlighted what can be called the 'official view' on gender on campus: the construction of young women's 'hyper-sexuality', the reluctance to engage with 'sexual harassment' even if members of the faculty are involved, the reliance on home (parents/guardians) for disciplining, and the pathologizing of 'aggressive' behaviour. This 'official' view of gender unfolded further over the many months that I spent on campus and has been discussed further in later chapters.

These official discourses on gender and sexuality are especially important in highlighting the relative absence of critical understandings of gender on college campuses, except at a few elite locations. The SAKSHAM Report (2013) published by the University Grants Commission also attests to the same. As a researcher of gender, it is important for me to juxtapose this 'lack' against some of the issues highlighted in the ethnographic chapters such as the surfeit of protectionist measures and attitudes, the view that students are 'hyper-sexual', and gendered interaction poses a 'risk', and the tacit acceptance of inequitable work cultures that limit female students from having the same access to opportunities as their male counterparts. No organized movement existed to counter or question any of the above, among the students or teachers. Even in private conversations, most members of the faculty and student body seemed compliant with the spirit of these rules; critiques were limited to the regimented nature of college life, or the high fees charged by private institutions. However, from time to time, violent protests have also followed certain incidents on private campuses. This includes suicides by students who cannot pay fees or backlash against college managements by those who feel humiliated by the harsh responses of college authorities to instances of 'indiscipline'. However, these are often quelled, and students 'suspended' or 'rusticated' before protests get out of hand.

Secondly, though one of my objectives was to study 'caste', it was an awkward and sensitive topic, because the colleges ostensibly followed a general policy of 'caste blindness' in the college, while adhering to regulations for caste-based reservation and disbursing various subsidies for students from various disadvantaged backgrounds. I was specifically asked by the college authorities not to directly ask anyone's *jati* identity or discuss these identities with students. Such discussions may lead to friction between the students, I was told. On the other hand, I was told that students were used to being referred to by the state-administrative categories such as OBC (Other Backward Class), Most Backward Class (MBC), Scheduled Caste (SC), and I was free to use these terms in my conversations for the purposes of my research. This peculiar paradox alludes to a larger issue persistent all over the country—the fall-out from the conflicting policies followed by the state regarding social justice and 'caste blindness' (Deshpande 2013). However, once I was entrenched in peer groups, it was obvious that most people in the peer group knew each other's caste identities despite the non-revelatory nature of Tamil surnames, but hardly referred to them, with the understanding that it was 'sensitive' (also see Nakassis 2016). Extended conversations with students, however, showed the pervasiveness of a view of caste as culture—young women referred to caste as a marker of an individual community's 'beauty', 'individuality', and 'aesthetics' within which kinship and other intimate relationships were nurtured (also see Natrajan 2012; Dumont 1980). Moreover, the idea of caste as a form of extended kinship also enjoys circulation through 'casted' friendship networks meant to 'look out for each other'. While aspects of this friendship include care and support, they could also be considered mechanisms of surveillance—conduits through which events at college were conveyed to family and clan outside, through which 'reputations' were built or marred (see Chapters 5 and 6 in this volume). This calls for expanding studies of caste to include aspects such as friendship and camaraderie, especially in a context in which many youths claim that 'friends are the new family' (Bhandari 2018).

Third, as I elaborate in the following chapters, class is inextricable from education. Class is rendered visible through several tangible and intangible factors such as the ability to speak English, the kind of primary and secondary schools students had attended, occupations of parents, and regional backgrounds. I would like to make a note here of the connections between class and the quota system. In addition to the caste-based reservations mentioned earlier, which are implemented both in the Government Quota and the Management Quota, the label 'Management Quota' is marked by class as it meant that students' families are financially capable of paying the hefty sums associated with management fees. Similarly, students coming through the Non Resident Indian quota are charged higher fees, are housed in a separate hostel with air-conditioned rooms, and served different food (which are also charged at higher rates). Invariably, the term 'NRI' has come to stand in for an elite class within the student body, with their lifestyles coded as elite. These 'NRI' linguistic and sartorial markers also carry moral connotations in the everyday, on which I will expand in Chapter 3.

Though religion is not an explicit theme in my work, I would like to add a short note on religion just as a matter of clarification. First, though Hinduism is the most widely professed religion (87.58 per cent), Tamil Nadu is home to significant populations of Christians (6.12 per cent) and Muslims (5.72 per cent) (2011 Census). Several colleges are run by the latter communities as minority institutions, for which they are granted separate quotas and other minority rights. Though many students enrol in colleges irrespective of denominational status of owners, there is greater minority representation in institutions run by their respective groups. These groups are well represented among the college owners as well as among the student body, as my study also reflects.

Second, in Tamil Nadu, caste is well integrated within the religious diversity, and cohesion within caste groups even among non-Hindus is strong (Hardgrave 1969; Ram 1991; Busby 2000). Christians, for instance, use a compound term comprising their religion as well as their caste names such as 'Latin Catholic Mukkuvar' or 'CSI (Church of South India) Nadar' to describe their affiliations. This can be attributed to the fact that conversion in Tamil Nadu has a long history and often the new divinities and teachings presented to the 'natives' by missionaries or brought back through trade were perceived and assimilated within existing cultural practices, 'little traditions', and the indigenous moral order (Bayly 1989). Perhaps this is why caste remains a tenacious feature of Tamil society in terms of occupation, inherited status, as well as maintaining endogamy, sometimes cutting across religious diversity of Hindu and Christian (Hardgrave 1969; Natarajan 2010). Thus, it is not even very remarkable to have different religious groups within the same family, or within a generation or two. This also means there is a commonality in cuisine and sartorial styles (these rights being hard won after a prolonged battle with brahminical Hinduism [Hardgrave 1969]), and Christians do not dress very differently from Hindus, as in other Christian communities in Goa, Mangalore, or the North East.

Islam also has a long history in Tamil Nadu, going back to maritime trade relationships with Southeast Asia, and shares many commonalities in cuisine and culture with the Indian Ocean world, rather than Islam as practised in North India or West Asia. Women of communities such as Rowther and Mayyakkara Muslims wear the same clothes that other Tamil women in the region wear, with perhaps the addition of a white *thuppatti*[30] garment similar to a *dupatta* draped around the head and shoulder for modesty. Tamil Muslims have a social stratification based on profession (Mines 1973) and some communities such as the Rowther Muslims have made huge strides in education, with their own education empires. These discussions take me to a third but related point that one should not assume an all-pervading Hindu culture in the southern Indian states such as Tamil Nadu and Kerala. As Susan Bayly argues (1989), religious communities do not have hardened boundaries in the region because of many shared customs, values, and practices. These two related features also explain the strength of mobilizations around Brahmin and Non-Brahmin dichotomies in the early twentieth century (Geetha and Rajadurai 1998), its diffuse form as 'Tamil' cultural nationalism by the mid-twentieth century (Barnett 1976), and the resistance against 'Hindutva' mobilization in the first half of the twenty-first century.

Organization of the BookAt the outset, I underlined the specificity of this youth study as one that aims to sociologically understand young people's dreams of mobility and aspirations to thrive and flourish as they manifest in the pursuit of a professional degree in engineering. These

---

[30] A length of cloth meant to serve as a modesty garment, similar to a dupatta.

processes are, of course, heterogeneous, fragmented, and differently imagined and pursued by everyone, subject to contingencies and change depending upon circumstances and constraints. If, for one, the pursuit of an engineering degree is the dream of securing a government job as a public works engineer, for another, it could be a back-up plan in case she does not pass the Civil Services Exam; for others, it could mean the prospect of successful marriage alliances, or the chance to expand the family businesses. Though myriad in form, these aspirations reflect the adoption of strategies by young people to surmount obstacles and achieve their goals of middle-class respectability, higher status, and financial stability. In fact, in the imagination of my interlocutors, life in engineering college emerges as the life-stage when the professional and the personal come together to interlock and propel them forwards. Thus, it encompasses aspirations of 'becoming professional' (Chapter 3) as well as intimate aspirations for finding a suitable mate (Chapter 5); it is the time of intense bonding with peers as much as being subject to the 'individuating tendencies' of surveillance technologies in the college (Chapter 4). Caste, gender, and class are centrally located in these issues, but also manifest in different thematics as listed earlier.

Given their range, I have organized the book into the following chapters: In Chapter 1, the metaphorical and conceptual frames presented by genealogy are used to establish the local histories of engineering colleges, through mechanization of agriculture and the region's participation in the global knitwear industry, bringing about prosperity and the necessary resources for investment in higher education. The chapter, thus, takes forward Richard and Rudnyckyj's (2009) argument that neoliberalism and privatization do not take place as a sweeping change, but move forward through the circulation of an 'economy of affect'. In the case of Kongunadu, successful privatization has been achieved through the secularization of traditional patronage as philanthropy as well as efforts by various backward class groups to enter education in general, and engineering education in particular. This has also sealed a new form of industry in the form of 'edupreneurship'.

In Chapter 2, arguing that 'institutional' big men (*periya aal*) have played a significant role in transforming higher education in Tamil Nadu through 'edupreneurship', I focus on their portrayals of power, their personal stamps on the process of education, and their characterization of education in familial and gendered terms. This chapter provides further context to the expansion in higher education by mapping how private engineering education has gone hand in hand with business, industrial capital, patronage, philanthropy, and electoral politics in urban and semi-urban milieus in Tamil Nadu. These details serve as necessary background reading to understand the ways in which the college serves as an institution of (im)mobility even as it fuels the aspiration for engineering education.

Chapter 3 marks a move from models of education to its contents (and discontents). Though engineering pedagogy remains outside the purview of this book, it is important to focus on how pedagogy has evolved to mould a new prototype of the private-sector engineer, fluent in English, American middle-class culture, as well as management activities. Training in this model has been adopted widely in engineering colleges under pressure to produce graduates who are 'employable', and not merely proficient in the manufacture and use of technology. Though students critique aspects of this training as being incompatible with local social mores (which the college enforces in everyday life), they also learn to focus on the self as the key locus of change. In such a process, existing class and caste inequalities are often reproduced and a person with existing signs of cultural and social capital are often determined to be more 'employable' than others, contributing to a widening gap between those who already have capital on their side and those who do not.

Chapter 4 shows how gender difference is reinforced on campus through language, through rules, regulations, surveillance, and control. I show how women enter and sustain their higher education *only* under conditions of being able to enhance their family's respectability. Private colleges, run by one's own caste association or a member of a similar caste, are able to enforce these conditions through various rules and regulations and surveillance of women in particular. Though women strategize to escape, reject, or negotiate the boundaries imposed on them, many of them are also forced to discontinue their studies when found to exceed the allowances made for them. Indian women in engineering colleges, therefore, face a different manifestation of the 'leaky pipeline' (Blickenstaff 2005) metaphor often used to describe women's high dropout rates from STEM careers. Unlike in the First World, where women are often considered academically unfit for STEM, in India it is the secondary status of women or the fear that they will engage in transgressive activities such as love, sex, and romance that emerge as major hurdles in the completion of their education.

If the chapters delineated above capture young people's aspiration for a profession, Chapter 5 attempts to capture students' personal lives—processes of courtship, selection of romantic and sexual partners, and marriage aspirations in the college, and its negotiation through social norms of caste and gender. This chapter is based on the premise that the space of an institution is not limited to technical education but is also an important aspect of young people's social lives as they romance, experiment with sexuality, and seek sexual/marriage partners. I pay special attention to students' articulations of caste and profession in relation to their choice of intimate partners, and to gendered subjectivities within heterosexual relationships. The choices made by young people illustrate the reproduction of certain class, caste, and gender orthodoxies, even as they attempt to achieve financial independence and higher education.

The concluding chapter revisits some of the key understandings arrived through extensive ethnographic research in engineering colleges in Tamil Nadu and shows the realities of higher education in contemporary India. Despite the manifold increase in the number of institutions and the subsequent entry of several first-generation learners, hierarchies and inequalities are largely reproduced. Women can enter education only under conditions that enhance their families' respectability and groups that do not have explicit markers of cultural and social capital are found to struggle to find employment and rely on local networks and brokers to find them jobs. These often involve large payments and further financial burdens on families. The extensive case studies can help us understand some of the seeming paradoxes that characterize contemporary India such as why despite a drastic increase in women's access to higher education, their enrolment in the workforce is at an all-time low; why young men and women struggle to find jobs despite having a professional education degree; and why marriage choices are often made along the grain of caste.

1
'Kinning' Education
Genealogies of Private Engineering Colleges

One November morning in 2018, I received a brown-paper package addressed to me with a thick book inside. Printed on acid-free paper, hardbound, and encased in a cover of ivory and gold, the embossed title read *The Fifth Brother: The History of a Noble Enterprise* (n.d.). It was sent to me by one of my interlocutors at the PSG[1] Institutions in Coimbatore in response to an enquiry on source material for the history of the institution. I had expected a long conversation, or at most, a short excerpt from the college prospectus, but had instead received this weighty tome. As subsequent pages in *The Fifth Brother* educate me, the title draws from the story that the trust was formed when four brothers, the sons of P. S. G. Rangaswamy Naidu, divided their ancestral properties not four ways but five, with the fifth portion invested in a charitable trust, thus 'kinning' their relationship. As I flipped through the pages, I dwelt on the image of the iconic architecture of the PSG Charities Building with its pointed arches, stained-glass windows, and twin minarets topped with cupolas and metal finials, which took me back to a campus visit in November 2014—though back then, I did not know the story of the 'brother'.

The metaphor is not a mere romanticization of the relationship between brothers as one that precludes trust and cohesiveness, providing a moral framework of familial ties within which the trust's affairs would be conducted, but is extended to every aspect of the trust's existence. Extended metaphor apart, the book draws me in as it weaves the history of the trust (and the engineering college that is my primary interest) as

---

[1] The abbreviation PSG that is used by the family as well as by their various institutions and industries stands for *Periyaveedu* (Big House) as well as two notable ancestors, Sama Naidu and Govinda Naidu (*The Fifth Brother*, 5).

the story of a family of Kamma Naidus from Andhra, a clan of warriors sworn to the Kakatiya rulers, who cultivated their lands with toil, became industrialists and patrons, and came to run a host of educational institutions. To introduce the family, the book describes clan lineages (*gothras*), local patron–client relationships, marriage alliances, and the family's activities as it is portrayed in eighteenth-century temple edicts. The history of the institution is then told through the life histories of pre-eminent men[2] in the family, their institutional affiliations, and the ways in which they enhanced the fortunes of the 'Fifth Brother'.

This story, which at its heart elaborates the intersubjectivity of institution and kin, becomes an invitation for me to enter histories of mercantile capital, regimes of labour, industry, patronage relations, and kinship even though my 'field'—engineering colleges in Tamil Nadu—is not a site in which one would traditionally start looking at these things. As I reflect on the life trajectories of generations of men in the PSG family who had played roles as '*dharmakartas*' (trustees) in local temples, started schools, and been at the helm of the family's industrial empire, I realized the engineering college as being embedded in familial ties is actually an important theoretical metaphor. In presenting the history of the institution through the metaphor of family history, the story of the Fifth Brother (unwittingly) points to the ways in which the methodology of genealogy is unavoidable for an intentional 'history of the present' (Foucault 1977).

Moreover, to those who know and understand South India, such invocations of lineage, charity, and trusteeship would not seem peculiar at all—such are the formulaic careers of 'big men', and portrayals of institutional affiliation of the kind described in the book are a common mode of introduction and self-presentation in these regions. They are meant to plot a person's place within a delineated set of institutional domains such as family, business, public service, institutional management, sponsorships, charity, clubs, and societies. These men are what Mattison Mines (1996) calls 'institutional' big men, whose lives are defined by the institutions they lead, practising a form of altruistic patronage that is highly valued in society, even while retaining a self-interest, and perhaps even

---

[2] It is important to highlight only men are visible in this volume, though indirect references suggest that women of the family were also actively involved in the education enterprise. See section on G. R. Govindarajulu (*The Fifth Brother*, 51), which says that his wife Chandrakanthi was a dedicated educationist. More details on Chandrakanthi Govindarajulu in the next chapter.

profit, at heart (also see Mines and Gourishankar 1990; Piliavsky 2014). They are credited with having increasing influence in India as 'education mediators' (Lynch 1990) and use their institutions to craft both 'influence' and 'social dominance' (Jeffery, Jeffery, and Jeffrey 2006; Alm 2010), intending to create particular kinds of young adults as well as bolster their own social, economic, and political image. They often appear as the sole 'benefactors' of particular neighbourhoods/towns, not only able to invite various local notables, people of national and international interest, and media attention to their little towns. In these ways, they wield considerable power, using their image as 'educationists' to contest elections, and even promising to open more institutions if they won seats (Jeffery, Jeffery, and Jeffrey 2006; Tamil Nadu case elaborated later in this chapter). These processes are fraught and hotly contested—especially whether the colleges can benefit the locality, or if it is only meant for profit and the glory of specific 'charitable trusts'. Yet this model has become immensely popular, especially for engineering colleges, resulting in the de facto privatization of the entire sector.

In this chapter, I try to capture the interplay between philanthropy, patronage, and profit that dictates the economic, social, and cultural logics of contemporary engineering colleges by excavating its historical lineages. This is an important (though by no means exhaustive) supplement to the histories of colonial policy on education, and nationalist struggle over *swadeshi* (indigenous) industry and teaching of skills, within which most scholars would locate the roots of contemporary engineering education (Basu 1982; Headrick 1988; Swaminathan 1992; Ramnath 2017). Relying on family memoirs, journalistic accounts, and other information available in the public domain, I highlight the 'personal' and 'social dimensions' of private engineering colleges in Tamil Nadu to give a sense of the milieu in which they are located, the discourses that circulate in them, and the 'economy of affect' (Richard and Rudnyckyj 2009) in which subjectivities are enmeshed. My objective is not to present a complete and exhaustive history, but to argue that privatization has been able to build on existing spheres of influence and is linked to other changes in the political-economic change in the late nineteenth and twentieth centuries.[3]

---

[3] Changes after the 1990s are encapsulated in the next chapter in this volume on management practices.

Though milestones such as Macaulay's *Minute on Education* (1835)[4] and the setting up of the Thomason College of Engineering in Roorkee in 1847[5] were important from the point of view of institutionalization and the formalization of technical education, contemporary private engineering colleges have genealogies linked to transnational mercantile capital, globalized flows, changes in agrarian relations, and commerce as much as local histories of family, caste, and Dravidian politics. This troubles the notion that engineering and technical education is essentially a British colonial legacy. Taking my cue from Aparajith Ramnath (2017) that there were multiple processes through which engineering became an 'Indian profession',[6] with expertise drawn from indigenous and Western practices as much as the construction of the profession through Indian institutions, I show the role of private capital in shaping technical education in India. Moreover, genealogy as metaphor and methodology emphasizes difference from this mainstream and developmental history and brings a cross-sectional perspective to change, their effects, the affect it has on stakeholders, and its implications. Terms such as 'kinning' show the power of the 'economy of affects' (Richard and Rudnyckyj 2009) in mediating the changed relationships wrought by neoliberalism. In this case, it locates higher education in a history of philanthropy and giving, which in India invariably has roots in caste and religion (Roohi 2016; Sundar 2013), but also intended to build individual and family stature (Mines 1996). My argument is that these aspects have to be placed at the heart of the privatization debate to truly understand how it has been able to encroach on philanthropy, riding on the idea that delivery of education in itself is a 'social good' irrespective of returns received.

---

[4] Following the directions of the Minute, universities were set up at Calcutta, Bombay, and Madras in 1857 for the 'natives'.

[5] In the mid- to late nineteenth century, engineering classes were also introduced in the universities of Madras, Poona, and Sibpur near Calcutta to train labour for the colonial government. High school training was also organized accordingly: a bifurcation of studies into two divisions in the upper classes of high schools was introduced in 1884; one leading to the entrance examination of the universities, and the other, of a more practical character, intended to fit youths for commercial or non-literary pursuits.

[6] Primary in his argument is the role played by private capital and large-scale industries such as the Tata Iron and Steel Company in training and recruiting its engineering personnel (Ramnath 2017, 183). Reading personal memoirs, company newsletters, and logs, Ramnath (2017) gives an exhaustive account of how the company was able to *Indianize* the profession by hiring American consultants, foreign-trained Indians as well as setting up its own training centres such as the Jamshedji Technical Institute (JTI) to 'mould personnel according to the company's needs' (2017, 207).

Patronage to Philanthropy: Technical Education up to the 1960s

The end of the nineteenth century and the beginning of the twentieth century saw education, especially technical and higher education, being politicized as it emerged as one of the bones of contention between nationalists and British authorities. Its introduction in India was dictated as much by colonial demands for cheap labour as the nationalist pressure on the colonial government to to provide better education to the natives. On one hand it was meant to supply technically trained personnel for the Public Works Department (PWD), and on the other it was also a means of 'supplying' education 'which many Indians saw as a means of national liberation' (Headrick 1988, 316). The nationalist P. N. Bose was among the first nationalists to raise the issue in a pamphlet titled *Technical and Scientific Education in Bengal* in 1886. In the document, he criticized the lack of industrial training offered in universities, decrying the focus on 'theoretical knowledge only' (cited in Headrick 1988, 328). He emphasized the need for training in dyeing, tanning, mining, soap, glass manufacture, sugar milling, and electrical engineering, and wrote that 'practical tests should form the distinctive feature of science examinations' (cited in Headrick 1988, 328). He also went on to advocate the founding of a Central Science and Technological Institute to encourage industry. Such a call was taken up by the Indian National Congress, which passed a resolution in 1887 'that having regard for the poverty of the people, it is desirable that the government be moved to elaborate a system of technical education' (cited in Headrick 1988, 329). Similar resolutions were also adopted in 1892, 1898, 1900, and were also echoed in the nationalist press (Headrick 1988, 329). Such demands were looked upon with disgust by the colonial government. Governor General Lord Curzon was convinced that Indians' demands for technical education were 'native clamourings for things about which they know nothing', and if they are educated, 'will only add to the discontented hordes' (Headrick 1988, 329). Though a few steps were taken to remedy the situation, the results, Headrick argues, were disappointing (1988, 330).

It was not until the Swadeshi movement in the early 1900s that technical education received a shot in the arm. Foreign goods were boycotted.

Sales of English cotton goods fell over seventy-five per cent, and the Indian cotton industry took this opportunity to expand the industry and raise prices. The movement's leadership also perceived a lack in education, and a National Council of Education was founded in 1905 to train Indians for the *swadeshi* industry. Several Indian industrialists also took up the cause, and the industrialist J. N. Tata, who was convinced that 'science was the handmaiden of industry', proposed to set up a scientific research institute that could compete with the best in the world. Lord Curzon, however, resisted the idea and it was not until 1909, after J. N. Tata died, that the Indian Institute of Science was set up, funded by the Maharaja of Mysore who also offered land for the campus in Bangalore (Headrick 1988, 336).

However, attitudes changed dramatically during the war years. During the First World War, Britain withdrew most of its engineers and equipment from India, and the Indian economy could not cope with the loss. Concern over India's weakness led to the appointment of the Indian Industrial Commission in 1916. The commission was very critical of the inadequacy of technical education and training in India and the policy of importing trained personnel, deploring the government's sponsor of 'literary and philosophic studies to the neglect of those of a more practical character'. Though critiques of this nature had been common, what was new was that the commission tied technical education to industrial development and gave the government responsibility for both. This new attitude gradually galvanized the government into action during the 1920s and 1930s, and the Indian government lumbered towards a policy of industrial development and greater representation of Indian personnel in various civil and engineering services. Five new engineering colleges and technical institutions were also established. After the war, more Indians went to England to study. New industries had also begun to spring up everywhere, and technical education had to evolve to keep pace. By the end of the Second World War, government records finally reflected that a change had come over the services: 'Many young men, who would otherwise not have embarked on a technical career, have been recruited under these schemes and the prejudice against industrial employment had been steadily breaking down' (*Report on Technical Education*, 1943, cited in Headrick 1988, 343). As Headrick sums up, 'a combination of rising demand, government support, Indianization, technical education and economic growth had broken the deadlock of colonial pessimism' (Headrick 1988, 344).[7]

Even as two world wars and the Great Depression brought sweeping changes in government policy regarding technical education and recruitment of Indian engineers, the history of the PSG family in this period tells the larger story of agrarian and industrial change as well as capitalist accumulation in Kongunadu. Cultivating about ten to fifteen acres (50,000–72,600 sq yards) of cotton and tobacco, the family benefited immensely from the network of canals from the River Noyyal, control of caste-based labour, and the ability to invest in mechanization of agriculture (Govindarajulu 2015).

As the First World War ravaged Europe and the Great Depression followed, the economy crashed and so did the traditional lines of credit that buoyed up the cotton trade in cities like Ahmedabad. In the economic restructuring that followed, cultivators began to initiate expansion into trade and processing, and towns in the cotton-rich areas such as Coimbatore and Tiruppur emerged as the new entrepôts of the cotton trade (Chari 2004a), with many new cotton mills set up in these locations in the early 1920s. The Sri PSG Ranga Vilas Ginning, Spinning, and Weaving Mills was one of them. This mill was instrumental in bringing electricity to Coimbatore from the nearby Pykara Hydroelectric Project, which in turn mechanized agriculture to a greater extent, resulting not only in a greater accumulation of resources for the family, but also the development of a small workshop for servicing the growing collection of machinery owned by the family's enterprises.

Between 1920 and 1926, this workshop too, evolved into an enterprise: the foundry began fabricating steel structural elements and manufactured castings needed for the cotton mills, and eventually, centrifugal pumps—a product that became the family's hallmark and revolutionized agriculture in the region (Zacharias 1950, 44–45 cited in Chari 2004a, 767). With rural electrification, more homesteads invested in electric pumps thus changing the agrarian landscape and consolidating regimes

---

[7] Using rich sources such as personal memoirs, government records, official reports, and trade journals, Aparajith Ramnath (2017) goes into detail about this period: focusing on the nature of industrialization or the growth of private industry between 1900 and 1947, and the subsequent 'Indianization'—the process by which more Indian engineers were inducted into government service and industry.

of casted and gendered labour through which dominant castes such as Gounder and Naidu landlords were able to not only enhance production, but also start small units for ginning, spinning, and weaving cloth lengths on their own (Chari 2004a).

As the PSG family's wealth grew from their myriad enterprises, their attention turned to philanthropy and the first school in Peelamedu, the family's stronghold in Coimbatore, was started for the poor in 1924. Intended to be more secular than the family's other patronage activities that included annual gifts to various temples, this marked the entry of the family from patronage in the traditional and religious sense to a secular and formal format as 'philanthropy'. Wanting to build on their religious patronage as well as the casted identity of Kamma Naidus as 'givers' (Roohi 2016), they also began a trust to manage the affairs of the school. The school itself was shaped by Gandhian ideals of 'giving back', uplift of the poor, and nation building through education for all. Accordingly, the school was named 'Sarvajana School' and characterized as a form of '*seva*' (service) for the poor. The school also worked closely with the foundry and the industrial enterprises. The ends of education and industry were perceived to be the same: the securing of a livelihood.

The PSG family's conceptualization of 'philanthropy' was different from other acts of philanthropy at that time such as that of Alagappa Chettiaar, another successful textile mill owner, who bequeathed land and money for a chain of institutions all over Tamil Nadu (including the Central Electro Chemical Research Institute at Karaikudi, and the Alagappa Chettiaar College of Technology in the present-day Anna University Campus at Guindy), whose control was firmly with the public sector.[8] Donations like Chettiaar's were the order of the day. Between 1892 and 1947, Indian philanthropy not only made the transition from merchant charity to organized, professional philanthropy, but did so on an impressive scale (Sundar 2013; Kapur and Mehta 2004). This period saw the establishment of some of India's most enduring trusts, foundations, and public institutions—Aligarh Muslim University, Banaras Hindu University, Jamia Millia, Annamalai, Indian Institute of Science, among others—were created largely through voluntary donations. What

---

[8] The pre-independence period, or that between 1892 and 1947, has been termed as the 'Golden Age of Indian Philanthropy' (Sundar 2013).

is even more striking, a major proportion of their grants went to 'public institutions' such as universities that were either directly under state control or some form of public authority. Although emanating from family trusts, these funds were not under the control of family trusts and were deployed for specific purposes by the terms set by the receiving institutions and not the trust itself. At the time of independence, the net share of private philanthropy in shouldering the burden of public institutions was as high as seventeen per cent in 1950 and is now down to less than two per cent (Kapur and Mehta 2004, 26).

However, the PSG family, and other industrialists such as C. Rajam (also a textile mill owner branching into steel) who started the Madras Institute of Technology in Chennai, were focused on the idea that the educational institutions they pioneered should be firmly enmeshed in their industrial interests; that students should spend time on the factory floor as much as in classrooms and labs (*The Fifth Brother*, 49). Thus, it is mentioned that P. S. G. Rangaswamy Naidu, who was the managing trustee between 1926 and 1947, saw the growth of the Sri Ranga Vilas Mills not just as an industrial enterprise but also as a place for innovation and experimentation, where education and regular work should go hand in hand, which has enabled it to remain an epitome of modernity through the years (*The Fifth Brother*, 49).

This principle influenced the subsequent establishment of institutions by the PSG family such as the PSG Charities' Industrial Institute (1926), the PSG Polytechnic College (1939), PSG College of Arts and Science (1947), and the PSG College of Technology (1951)—a relationship described through the metaphor of cross-fertilization because of the industrial experience acquired by students as well as the industry's benefit from the in-house research and development (R&D) of faculty members.

The PSG College of Technology was a dream project of P. S. G. Rangaswamy Naidu and his son, G. Damodaran Naidu (GRD), who went to the UK for higher studies in engineering. The story goes that in the UK, GRD and his peers from Coimbatore, the brothers G. K. Sundaram and G. K. Devarajulu (from the family that owned Lakshmi Mills), and K. Srinivasan envisioned a collective dream to turn Coimbatore into the 'Manchester of India'. Whereas the brothers intended to start an industry to manufacture machinery for the textile industry, GRD promised to start an engineering college that would 'produce qualified manpower to take care of the needs of his friends' industries' (*The Fifth Brother*, 105). K. Srinivasan started the South Indian Textile Research Association (SITRA). Their friendship thus embodied a productive relationship between education, research, and industry. GRD, who graduated in Mechanical and Electrical Engineering, came back from the UK and took over as the director of the PSG Industrial Institute (PSGII), and started the groundwork to begin an arts and science as well as engineering college, which finally opened in 1949 and 1951 respectively, affiliated to the University of Madras.

Similarly, another well-known industrial in Madras, C. Rajam started MIT with the Rs 0.5 million (about $6,000 US) he received from selling his house, after finding that Indian engineers were inadequately trained to install and run the imported machinery at his new steel-rolling plant (Rajan 2014). As was the situation at the Tata Iron and Steel Company (TISCO) (Ramnath 2017), Indians did not possess the requisite know-how to operate machinery imported from the West, and sending personnel to the West for education was an expensive proposition. In the face of these obstacles, starting an institution to train personnel remained the only feasible solution if the steel plant was to function, though its motto said, 'In Service of India'. MIT was set up in 1949, even before the first Indian Institute of Technology, and offered engineering degrees in four streams, with each department headed by professors from Germany.

On the website of the Madras Institute of Technology (n.d.), the event is described thus:

> In 1949, Shri. C. Rajam gave the newly independent India Madras Institute of Technology, so that MIT could establish the strong technical base it needed to take its place in the world. It was the rare genius and daring of its founder that made MIT offer courses like Aeronautical Engineering, Automobile Engineering, Electronics Engineering, and Instrument Technology for the first time in our country.

To suggest that some self-interest lay behind engineering colleges is not to discredit these captains of industry—they were truly motivated by the idea of 'giving' to the new nation and made many personal sacrifices on their part. Rajam was among the industrialists who were heavily influenced by Mahatma Gandhi's principles. Rajam's bungalow, 'India House' (which he sold to start MIT) is known to have been extremely ostentatious, and the PSG Charitable Trust is said to have absorbed several losses in its folds in order to introduce new trainings in the workshops as well as in the making of several prototypes that were never realized (*The Fifth Brother*, 51). Pushpa Sundar (2013) points to a combination of factors why the business community was so inspired to *give*; reasons that range from the entrepreneurs value of frugality and their belief in the virtues of education, to the emergence of vast industrial empires, and Gandhi's ability to encourage nationalist sentiments among the business class. As a result, 'earlier, money had been dispensed through trusts to meet immediate social needs, but the new trusts began to view giving as an instrument of social change' (Sundar 2013, 137).

These discourses also highlight the turn to technical education supported by private industry, and the gradual strengthening and 'remedying' of British-introduced English education which had been perceived as being 'too exclusively directed to university studies, and that no opportunity is offered for the development of what corresponds to the "modern side" of schools in Europe' (Report of the Education Commission (1983) cited in Swaminathan 1992, 1662). Just as TISCO obtained American expertise (Ramnath 2017), places like MIT were able to recruit faculty members from Germany, which was undergoing reconstruction under the Marshall Plan—alternate centres from which India was able to consolidate knowledge of engineering.

Though these contributions marked what Pushpa Sundar (2013) calls the 'golden age of Indian philanthropy' there were also negative implications: urban and caste bias, the use of unethical business tricks, excess profiteering, or a lack of concern for the natural environment. In Tamil Nadu, private engineering colleges and their upper-caste ownership perhaps deepened existing chasms between Brahmin and non-Brahmin categories around which there was considerable racial and colonial bias in the late nineteenth and early twentieth centuries. These were motivated around concerns of employment and education: Brahmins were favoured for employment by the British, and their absorption as engineers into departments such as Public Works Departments resulted in considerable ire against their 'domination' fuelling support for non-Brahmin politics between the years 1917 and 1929 (Fuller and Narasimhan 2008a, 179–180). This resulted not only in political agitations such as the non-Brahmin movement, but also strategic deployment of resources for the mobility of specific castes and groups.

In the Kongunadu area, where this study is based, the dominant agricultural group called the Kongu Vellala Gounders (henceforth called 'Gounders') organized into mobilized associations, both to control local communities and to access wider opportunities, particularly from the state (Arnold 1974).[9] This was part of a wider movement when 'caste began to secularize itself" (Washbrook 1975, 192) to access new political resources, developing a new idiom through which caste was pliable enough to allow patronage networks extending from the countryside to colonial institutions, and included political reservation that opened opportunities not just for the rural elite, but also for many of the poor and the rural dispossessed of Tamil Nadu (Irschick 1986).

However, independence saw a temporary lull in such activities and agitations, and mobilizations around caste divisions were briefly subsumed within the language of nation building and cultivating 'a rational and scientific temper' as envisioned in the constitution. This vocabulary dominated both policy debates and political discourse in India, conjuring up an 'Indian darkness' in 'need of science' (Roy 2007, 105–132). According to such a discourse, India was defined by the agents of the postcolonial state as a 'collection of persistent and unfulfilled problems, failures, and needs' (Roy 2007, 106). In articulating a vision for the future, this 'needs discourse', Srirupa Roy says (2007, 106), projected science as the need of the hour. Policy debates and political discourse stressed that India needed science to face its many problems in the economic, social, educational, and cultural arenas. The Radhakrishnan Commission was appointed to investigate the reconstruction of university education for meeting the scientific and technical manpower requirements of the country. The commission noted that for 'a fuller realization of the democratic principles of justice and freedom for all, we need growth in science and technology' (cited in Kaul 1993, 33). Several initiatives for the development of scientific and technological education were also taken; elite

---

[9] Such efforts were made with much success among groups such as the Nadars. This led to many castes. Robert Hardgrave's study (1969) documents the uprising among the Nadars against social oppression. The Nadar Mahajana Sangam, a caste-based association, promoted educational activity, welfare, and philanthropic work amongst the Nadars in the early 1920s, leading to greater levels of education in the community.

institutions such as the Indian Institutes of Technology were important and much-publicized priorities for the state; measures were also taken to set up postgraduate institutions, industrial research centres, and research laboratories dedicated to the pursuit of scientific research, in different parts of the country. Roy (2007) argues that located in unknown and unfamiliar 'elsewheres' (such as Karaikudi where Alagappa Chettiaar had offered 300 acres (about 0.5 square miles) for the Central Electrochemical Research Institute (CECRI) [see above]), these institutions enabled the reimagination of national space along a new cartography of science.

In such a milieu, initiatives such as C. Rajam's MIT were seen as fulfilling a primary need for the nascent nation, spurred by the 'needs discourse' (Roy 2007). These men received plaudits from national leaders like Nehru, who nicknamed the 'swadeshi industrialists' as 'socialist capitalists', and considered them to be the need of the hour (Rajan 2015). In regional parlance, they were '*periya aal*', who were already well known in the region, with significant amounts of social, economic, and cultural capital, who were known to cultivate their constituencies (Mines 1996). Starting engineering colleges not only brought them political mileage, cultivating a positive relationship with modernity and technology was also an important mode of portraying themselves as 'big men' with status (Osella and Osella 2006, 53–76). Entry into these few colleges was marked by intense competition, which sealed the status of engineering as being truly 'meritocratic' and over time became the preserve of those who had the requisite social and cultural capital, a category that mostly intersected with 'caste' Hinduism if not Brahminism, enabling the educational mobility of groups such as the Brahmins,[10] Naidus, and the Chettiars. These groups had considerable rural dominance and were able to invest in education and advance their way into institutions of higher education, and eventually into bureaucracy, banking, and the commercial private sectors

---

[10] This, of course, has a longer history: Between 1917 and 1929, there emerged a strong critique of the brahminical domination of administrative posts. The Brahmins ascendance in the ranks of employment and their inclination to make use of educational opportunities resulted in an urbanizing trend, a factor that was critical to their dominance of administrative jobs. This dominance led to a 'conflict between the landowning non-Brahmin elite with a history of rural dominance, and the nascent urban Brahmin elite that had used the opportunities presented by British rule' (Barnett 1976, 17). The non-Brahmin movement that criticized this domination and the Justice Party (formally titled the South Indian Liberation Federation) tried to position Brahmins as cultural colonists in Dravidian lands, undercutting Brahmin political power and access to British colonial bureaucracy.

as a result (Fuller and Narasimhan 2008a, 179–180; Subramanian 2015a). The visibility gained by these groups reinforced the imagination that careers in science, technology, and engineering became the prime means of social advancement (Beteille 1991).

This discourse was also instrumental in making engineering the most important profession, as it would involve men directly in the task of nation building. Hence, engineers gained a renewed sense of respect at the helm of Indian economic and industrial development post-independence, benefiting from Prime Minister Nehru's import substitution policy. Moreover, the idea that engineering is synonymous with social mobility permeated deeply into the middle class in the first half century of the postcolonial nation, and entry into engineering colleges was instrumental in professionalizing and urbanizing the middle class. Many who shaped the next few decades of Indian history were also cultivated at these institutions, lending them further legitimacy.[11]

Consolidation of the Private Model: 1970s to 1990s

Despite the emphasis on education for technical advancement, education overall remained a low priority going by budget allocations. Fund allocation for education from the First Five-Year Plan to the Sixth Five-Year Plan shows a steady decline (*The Fifth Brother*, 49), even though the middle-class aspiration and demand for higher education surged. This marked a shift from the 'golden years' post-independence to what Pushpa Sundar calls the 'winter of discontent' (2013, 163). These were difficult years economically, and the same was reflected in the engagement with philanthropy by private enterprises. The growing mistrust between business and government, increasing regulation, high taxation, and the more materialistic culture of a new generation of business owners resulted in a deceleration of the kind of philanthropic activities that poured money into the state coffers for education. This paved the path for the 'third sector'—a space of social action between the state and the market

---

[11] Former President of India Dr A. P. J. Abdul Kalam graduated from MIT with a degree in Aeronautical Engineering and Shiv Nadar, who set up HCL Technologies, graduated from PSG Technology.

to make inroads into education through acts of 'philanthropy'. However, in contrast to rational, targeted, and professional forms of giving which marks philanthropy in countries like the US (Sundar 2013; Roohi 2016), philanthropy in India has emerged as a cover for private enterprises to take over the affairs of the state, marking what Kapur and Mehta (2004) call the 'privatization of philanthropy'.

As Gandhian idealism and the Nehruvian vision of a country united by science and technology ebbed by the 1970s and 1980s, the 'cartography of science' spread across the entire nation state was displaced into atomized units of region, sect, and community. Philanthropy began to be directed inwards, taking the shape of securing class mobility for individual communities. With donations from the community, colleges and hostels were started for the benefit of the communities as admission to the handful of government-run colleges was found to be difficult given the competitive milieu and historic advantages of some communities (Subramanian 2015a). However, the links between industry and 'political entrepreneurs' (Wyatt 2010) in electoral politics was useful for communities to secure greater representation in affirmative action policies.

In Kongunadu, Kovai Chezhiyan, an industrialist and prominent politician in the Dravidian party from the Gounder community, was instrumental in getting Gounders included in the list of OBC (Other Backward Class) castes in 1975, despite their 'dominance' (Beck 1972) in the region through land ownership and control over regimes of caste-based labour (Chari 2004a). Chezhiyan also floated the Kongu Vellala Goundergal Peravai, a caste-based association, and often said in his speeches that education is the 'single most important asset that parents can bequeath their children' (KNMK 2009 cited in Vijayabaskar and Wyatt 2013, 106).

The Kongu College of Engineering was also established in 1983, by the coming together of '41 philanthropists from different walks of life who realised the need for technical education for their region's economic welfare, formed collectively a Trust called the "The Kongu Vellalar[12] Institute of Technology Trust" and they tried to promote and develop equality of opportunity for the rich and the poor' (Kongu College of Engineering website, n.d.). The website of the institution not only makes clear the connection between the 'Kongu' region and the community of

---

[12] Kongu Vellalar is the formal name for the Gounders.

Gounders, but also speaks of the erstwhile glory of the Gounders, their current ('depleted') status, and underlines the need for engineering education for the development of the region (Kongu College of Engineering website, n.d.).

The website's equation of caste with region signals the ways in which caste not only attained a secular identity and reinforced itself in the urban, from *jati* to *samaj* (society) (Dumont 1980, 221; Natrajan 2012), but also reveals the processes of class restructuring.[13] Class formation, historian Christopher Bayly argues in the context of north India, was tied up with the reworking of caste and community so that caste communities became 'mobilized associations for the defence of property, trade, and status' (1992, 481). Sharad Chari (1997) argues that the rise of the Gounder elites in Tamil Nadu, at least in part, resulted from their organization into such 'mobilised associations ...both to control local communities and to access wider opportunities in the state' (Chari 1997, 70). The case of the Kongu College of Engineering shows the effectiveness of these systems in arenas such as education.

This was not an isolated case. Several such trusts, with religious-, sect-, and caste-affiliations began to take control of engineering education all over South India in the 1980s (Kaul 1993; Ambrose 1994; Upadhya 1997). Along with caste associations, several religious sects, which attracted (and still do) magnanimous donations from the middle-class Indians were able to make inroads into education. Many of these sects are spearheaded by a Hindu monastic orders (*mathas*) and there are thousands of *mathas* in South India with practically all castes from 'dominant', backward, and Dalit castes having found it necessary to establish *mathas* of their own. The historian Janaki Nair quoting M. M. Kalburgi notes that the non-Brahmin *mathas*, some of which have existed since the fifteenth century, are *samajamukhi*[14] and engage the social in a wide variety of ways, including providing modern professional education (Giriprakash

---

[13] Similar efforts can be seen in Karnataka in the 1990s, when associations such as the Karnataka Lingayat Society were formed (Kaul 1993, 38–83).

[14] Brahmin *mathas* are generally considered to be *dharmamukhi* though some prominent Brahmin *mathas*, in the late twentieth century, have realigned themselves to the social by being actively involved in politics (such as involvement in Babri Masjid demolition), starting schools and colleges, protesting against certain kinds of developmental activities. These include the late Jayendra Swamiji of the Kanchi Kamakoti Trust and the Pejawar Matha seer in Udupi, Karnataka.

2018).¹⁵ In states like Karnataka, the various non-Brahmin *mathas* have literally been responsible for providing high levels of technical education. In the1990s, the heads of five *mathas* (*mathadipati*) contested a court case in Karnataka seeking fifty per cent of places at colleges as 'capitation fee seats' ('management seats') and were even ready to face arrest for the same (Pinto 1994).

This has also been the trend in Tamil Nadu, albeit on a lesser scale; various religious heads have been able to successfully cultivate their image as 'institutional' big men, with professional education forming a significant part of their contribution to society (Mines and Gourishankar 1990). These include personages such as the Brahmin seer of the Kanchi Kamakoti Trust, Jayendra Swami (also see Mines and Gourishankar 1990), and the non-Brahmin 'saints' Thirumuruga Kirupananda Variyar of the Vinayaka Mission Trust and 'Amma' Amritanandamayi of the Mata Amritanandamayi Math, whose trusts are at the helm of veritable 'education empires' today. Having cultivated an enormous following in India and abroad, the trusts are the receptacles of thousands of dollars from devotees. This money has been used to initiate several philanthropic as well as for-profit initiatives, supposedly intended to plough money back into the trust, sparking off controversies about the nature of what passes as philanthropy today (Zacharias 2008). The colleges also have a management quota, through which seats can be accessed through the payment of a donation to the trust, with each place (especially postgraduate medical seats) going for more than a crore of rupees (120,000 USD).

Moreover, these figures are also credited with the resurgence of Hinduism in the state after the rationalist movement lost its influence, and they have become influential members of the Sangh Parivar. Though invested in providing secular professional education open to all, one must consider the wider implications of the sphere of influence within which these personalities are rooted, and the contributions they have made in boosting the image of the nation as a 'New India' rooted in its traditional past, even as it can leapfrog stages of development into a globalized

---

[15] In a report in *The Hindu Businessline*, K. Giriprakash recounts how the seers became powerful after the former Chief Minister Ramakrishna Hegde granted them land to start engineering and medical colleges.

technical prosperity through the middle-class espousal of professions such as engineering (Nisbett 2007).

In fact, religious fundamentalism of all colours and hues have been able to ride this image of middle-class values leading to economic prosperity. Take the example of the Christian ministers and tele-evangelists Brother D. G. S. Dhinakaran and his son, Paul Dhinakaran, who started the Karunya Technological Institute in 1986. The institution's founding has been described on their website (Karunya University 'About Us' page, n.d.) as:

> He (Brother D. G. S. Dhinakaran) started Karunya Institute of Technology on 4 October 1986 in spite of losing his only daughter in a tragic road accident on his way to Karunya on 21 May 1986. His vision for Karunya is that of a Technological Institution founded on faith, which will produce teachers, engineers and managers possessing the right combination of academic excellence, exemplary character and total humanism. They should serve the motherland and fellowmen and help raise the quality of life to global standards.

It should be noted that Brother Dhinakaran's trust is not a missionary organization in the traditional sense but is better understood as a 'family business' (Thomas 2008, 121) run in nexus with the 'Jesus Calls' ministry and 'God TV' channel. The 'global Christianity' messages preached through their tele-evangelism imbricates with the prosperity-driven discourse of economic globalization, that 'the pursuit of wealth and profit making are premier virtues, over and above the pursuit of inclusive, communitarian futures' (Thomas 2008, 121).

In fact, both resurgent Hinduism in the form of New-Age gurus such as Jagadish Vasudev ('Sadhguru' Jaggi Vasudev) (Upadhya 2013) as well as 'global' forms of Christianity that rely on the fruits of globalization to deliver the message that Christ bestows health and wealth exclusively to his followers (Thomas 2008, 122) have one thing in common: they represent the reworking of religion into something that appeals to software professionals and other white-collar workers (Rudnyckyj 2010; Upadhya 2013), and by extension, to engineering students working on creating desirable 'employable' subjectivities. Whether it is Vasudev's 'inner engineering' or the reworking of Christian hymns into 'rock music' played at college events, they highlight the reassertion of middle-class family values as well as the development of desirable 'entrepreneur-worker' subjectivities. Though I highlight in Chapter 3 that the desired worker-subjectivities are not always realized, it should be underlined that institutions do not see any dissonance in the teaching of science and technology along with tenets of religion, often of a very fundamentalist strain, because of the common threads of globalism and capitalism in which both are shrouded.

The third kind of ownership I would like to highlight is of 'entrepreneurial politicians' (Wyatt 2020), who began to focus on developing educational institutions as a distinct mode of capturing political and social power. Even as certain groups were able to recognize the importance of technology for progress and mobilize their resources accordingly, it has also been argued that the rational–scientific Nehruvian discourse left a great mass of the Tamil population alienated, a void that was eventually filled by the rhetorical thrust of the Dravidian movement and its stalwarts, Canjeevram Natarajan Annadurai and Maruthur Gopalan Ramachandran (M.G. Ramachandran) (Subramanian 1999, 157).[16] Several policy interventions by former Chief Minister M. G. Ramachandran (hereafter MGR) were crucial to the shift in power from state to private sector and in aiding the mobility of middle castes. Among them, was a policy to introduce engineering colleges to cash in on the social imaginaries that equates an engineering degree with white-collar employment, upward social mobility, and migration. Many of these were started in the 1980s, when a set of 'educational mediators' were

---

[16] A protégé of 'Periyar' Erode Venkatappa Ramasamy (EVR), 'Arignar' Anna (The Learned Elder Brother as C. N. Annadurai was known), broke away from EVR's Dravida Kazhagam (Dravidian Federation) to pursue a more moderate political philosophy and started a new electoral party called the Dravida Munnetra Kazhagam (Dravidian Progressive Federation) in 1949. The DMK gradually shifted its propaganda strategy from the populist struggle to undermine local Brahmin power to a broader ethnic battle pitting Tamil Nadu against the power of India's federal government. Language was used as a political tool, and oratorical skills especially in pure Tamil were greatly prized and replicated on the cinema screen (Pandian 1992). After Annadurai's death in 1969, M. Karunanidhi succeeded him as chief minister and leader of the DMK. Although M. Karunanidhi and MGR began as close associates, a rift soon emerged between the two as Karunanidhi felt threatened by MGR's popularity as a screen idol and eventually dismissed him from the party. MGR began his own party in 1972 named after his mentor—the Annadurai Dravida Munnetra Kazhagam (ADMK)—staking a claim as his rightful heir. In 1976, the ADMK was renamed the All India Annadurai Dravida Munnetra Kazhagam (AIADMK). In 1977, just five years after its formation, the AIADMK defeated the DMK, and MGR bore the mantle of chief minister.

constituted by the educational policies adopted by MGR after an election promise by him to introduce more engineering colleges.[17] After they won in their specific constituencies in the 1984 polls, land was granted to party loyalists such as G. Vishwanathan, S. Jagathrakshakan, and A. C. Shanmugam[18] to start private engineering colleges. They were chosen with the intention of enabling the mobility of middle castes in Tamil Nadu such as the Nadars, the Gounders, the Thevars, and the Vanniyars.[19] In total, sixteen colleges were started during this time. In an interview published by *The Times of India* (Nurullah 2016), Minister for Education in the MGR Cabinet, C. Aranganayagam recalled the logic behind such a strategy, 'I strongly believed that TN could not advance without the progress of its caste groups. We invited S. Jagathrakshakan and G. Viswanathan to start colleges for the benefit of the Vanniyar community. Soon after, we gave licences to Jeppiaar[20] (of the Mukkuvars[21]) and A. C. Shanmugam (of the Vellalla Mudaliars[22])'.

MGR combined this move with another paternalistic gesture: the abolition of the Pre-University College (PUC) and the inclusion of classes

---

[17] MGR is known to have mobilized a similar model for medical colleges as well. In 1985, PSG Institute of Medical Sciences and Research was started by G. Varadarajan, Management Trustee of the PSG Trust, with the 'active support of his friend and the then Chief Minister MGR' (*The Fifth Brother* n.d., 60).

[18] G. Viswanathan started the Vellore Institute of Technology whereas the S. Jagathrakshakan began the Bharath Institute of Science and Technology (now BIHER). S. Jagathrakshakan was MGR's biographer who later started his own party, Veera Vanniyar Peravai and was also Union Minister of State for Commerce and Industries. A. C. Shanmugam runs the Dr MGR Educational and Research Institute. He also won the 1984 parliamentary election for the AIADMK, but has now started his own party the Puthiya Needhi Katchi (the New Justice Party).

[19] Vanniyars is the preferred name of the Vanniyakula Kshatriyas, a traditional agrarian caste. They are also referred to as Padaiyachi, Agnikula Kshatriya, and Palli. In the late nineteenth and early twentieth centuries, they organized to declare themselves as not a low caste. In the late twentieth century, under the leadership of Dr S. Ramadoss, they demanded to be declared as a most backward caste group (MBC) (Singh 1997, 1562–1565).

[20] Jesuadimai Pangu Raj is popularly known by his initials J. P. R., but in a spelt-out format as Jeppiaar. Discussion follows in the next section. He was at the helm of the Jeppiaar Group of Institutions.

[21] The fisher caste of Mukkuvars or Mukkuvans are distributed mainly in the southern districts of Kanyakumari and Tirunelveli. Most of them are Latin Catholic, sometimes referred to as Roman Catholic, and have the constitutional status of backward class (Singh 1997, 848).

[22] The Vellala Mudaliars are a well-known and well-distributed community in Tamil Nadu. They consider themselves to be the most superior non-Brahmin group, next only to the Brahmins in caste hierarchy. Originally, landowners and peasants, they were employed by the British in key positions in government service, and engaged in trade. They have also migrated widely from rural to urban areas, taking up jobs in educational institutions, in government and private organizations, and in business. A large number of them are in high-status modern professions such as academics, medicine, law, and industry (Singh 1997, 1670–1675).

eleven and twelve in higher secondary education thereby widening the net from the 200 colleges that offered PUC to 5000 schools that could do so. Many rural students would not have to travel far for classes eleven and twelve; it could be obtained from the nearby higher secondary school itself (Nurullah 2016). Thus, more students would be qualified to enter higher education. For MGR, this was a socialist move, meant to be a great leveller.

On the one hand, such a move helped MGR consolidate his hegemony within the party cadre. On the other, it fell in line with Dravidian politics, for which education had always been a site of struggle on the basis that Brahmins had held on to power in the state through their control over education. Following the politics of 'Periyar' E. V. Ramasamy and the development of literacy among the lower classes, DMK politicians such as C. N. Annadurai were often depicted in smaller villages and towns across Tamil Nadu as learned men, holding a book (Pandian 1992, 49). Moreover, MGR, in his own films, had dominantly portrayed himself as literate even if he was playing lower-class figures such as a cowherd or fisherman. M. S. S. Pandian (1992) argues that this was part of the DMK's strategy to emphasize that only education and literacy could emerge as a weapon against oppression by the elite. Following the idea that MGR's cinema and politics can be read as 'co-texts' where the image, metaphors, ideas. and ideology created in the cinema played a role in political success (Pandian 1992; Rajanayagan 2015), it can be argued that the move to announce more engineering colleges in the state sought to co-opt many dialogues from his films that tacitly espoused education for all.[23] With the momentum earned by this rhetoric, both Dravidian and cinematic, and the social context in the state where engineering education could only be accessed by a select few who had the cultural capital to do so, the election promise to introduce more engineering colleges could not have been more alluring.

---

[23] For example, in the movie *Rickshawkaaran* (1971), MGR can quote Shakespeare to a judge and say that 'An occupation can earn respect and honour, only if it combines physical labour and knowledge' (quoted in Rajanayagam 2015, 70). In another film *Kumari Kottam* (1971), MGR stresses the importance of 'earning while studying'; in *Panakkara Kudumbam* (1964), he says 'Man needs education to satiate his intellectual hunger and food to satisfy the physical hunger' (cited in Rajanayagam 2015, 70).

Such analysis, which is concretely tied to certain decisions and the traction they have within a local culture, is very important in supplementing aspects of leadership such as Weberian charisma (Weber 1947) which have been used to explain MGR's effectiveness as a leader (Pandian 1992). Mines' (1996) work is important in bringing the relational nature of leadership to the forefront, as is Pamela Price's (1989) work, which shows that 'person-centred aspects of Indian political behaviour' (1989, 559) needs to be explained with reference to local culture. Sketching the network of colleges started by MGR's acolytes shows the pre-eminence of MGR as a 'big man' and how he is linked 'to all members of his group' (Mines and Gourishankar 1990, 764) who form 'bridge leaders' (Mines and Gourishankar 1990, 773).

Such emerging patterns of ownership were formalized with the New Education Policy of 1986, which underlined the need for restructuring the education system to meet the challenges of emerging areas of science, technology, and modernization. This involved encouraging private entrepreneurs to contribute to educational development. The National Policy on Education (1986) also stated that resources could be raised by mobilizing donations and 'by raising fees at higher levels of education'. Kapur and Mehta (2004) argue that the government is also considerably misguided on what constitutes philanthropy versus what is investment. Because of its ideological commitment to state support for education on paper with allowances made for philanthropy, the state requires all trusts to be registered as 'non-profit'. However, 'non-profit' has been differently interpreted by various stakeholders: from being run by family trusts who charge no fees for students to colleges which charge students to be 'financially sustainable', or use colleges as a 'cover' for business interests (see next chapter). Moreover, trusts are allowed to receive 'donations' which are exempt from taxes, and private engineering colleges have perfected the system of formally collecting 'donations' as 'management fees' for places that come under the 'management quota'.[24]

---

[24] During fieldwork, it was observed that admission for management quota started even before class twelve results were declared, and 'cut off' marks were fixed based on pre-board exams conducted by schools. The 'cut off' list was prominently displayed outside the office of the Dean of Admissions. The area was policed by security guards who monitored that no one click pictures or carry mobile phones or cameras into the office.

These practices also beg the questions of who gains control over education through 'donations', in what ways do they exert their influence on everyday running of colleges and how does this impact education itself? Does it really allow for better control of outcomes from educational activities, or does it enable only some groups to monopolize control over resources by augmenting their political power? Kapur and Mehta (2004) argue that the 'privatization of philanthropy' has adverse consequences for the credibility of public institutions and philanthropic activity to higher education in general because they tend to 'encroach' on public institutions, and begin to serve ideological and vested interests. Sidel (2002) points out that they tend to enhance pre-existing fractures in society, particularly in the case of funds channelled to religious or political organizations. In the case of colleges, I argue they tend to act as 'vectors of control' that replicate forms of sociality based on kinship and caste networks, effectively acting like closed moral economies of the kind seen in transnational Tamil communities (Wise and Velayutham 2005).[25]

The IT boom and Beyond: 1990s to the Present

By the 1990s, the middle class expanded to the effects of education, professionalization, trade, and commerce, and the 'dream' presented by liberalization settled around the nascent Information Technology industry. The then Prime Minister Rajiv Gandhi spoke of the promise of modernity and technology, and of moving India into the twenty-first century (Nisbett 2009, 30). Such discourses especially held sway because this was the time that various specializations were also emerging within the engineering discipline. The newer disciplines such as 'software' and 'information technology' were not just premised on India's development and progress but made careers subject to global market forces. This was enabled by the 1984 IT Policy, the arrival of several American companies, as well as a nascent Indian IT industry that began to exploit the returns

---

[25] In their article, Wise and Velayutham (2005) show how the original inhabitants of Soorapallam village in Tamil Nadu, though now living in Singapore, continue to maintain social and cultural ties to their village, and base their everyday practices on a moral economy of obligations and responsibilities based on caste membership, which, in turn is regulated by regimes of affect and policed through the gaze of fellow translocals.

offered up by the offshore model. The resultant boom was the result of the relocation of IT services from the US to cities in India to embrace a cost-effective model in which Indian personnel would interact with their clients on-site to fulfil their requirements. The resulting culture of doing business has been widely documented (Upadhya and Vasavi 2006; Saith and Vijayabaskar 2005; Nisbett 2009; Upadhya 2016a) with the local recruitment of engineering graduates as IT workers emerging as one of the key highlights.

For software companies, the recruitment of engineering graduates as IT workers promised a workforce fulfilling requirements such as the technical competency required to learn coding and programming as well as English and communication skills. 'Engineers' as a category of professionals, were also more likely to be granted business visas when they had to travel to the US or Europe, to work 'on-site'. Moreover, engineering institutes were known to attract the best talent, given the durability of the Nehruvian vision of India as a great scientific technocracy. Companies were lured by the 'merit' of students from prestigious state-funded schools such as the Indian Institutes of Technology (IIT) as well as private engineering colleges known to churn out thousands of engineering graduates every year. Indian engineers emerged as a force to be reckoned with on the world stage, making up a highly valued set of white-collar workers in the US and elsewhere and paving the path for mobility and migration.

This set the scene for the decade following liberalization, which witnessed growth in the Indian IT Industry on a scale hitherto unknown (Nisbett 2009, 26–47), a growth that was possible only because of the proliferation of engineering education that went concomitant to it (Kamat 2011). This was also actively supported by the government which invested heavily in telecommunications, Internet infrastructure, and creation of software export-focused business parks (also in Upadhya and Vasavi 2006, 9–13), while companies attempted to demonstrate that Indian workers could meet global standards of productivity and efficiency. Such a climate also offered the chance for India's burgeoning middle class to participate and engage in what is referred to as the new middle class (NMC) culture; for young people, this could be achieved through access to employment in the IT sector. Education, especially engineering education, became a central social value for the middle classes as well as an important form of cultural capital which allows it to consolidate its social power, reinforced by the discourse of the knowledge society.

Nicholas Nisbett (2009), in delineating the celebratory middle-class discourses of 'the knowledge society' in India, says that at the core of the discourse is the assertion that India can 'leapfrog' stages of large-scale industrialization to a postindustrial, service-dependent economy. This narrative has also received significant impetus from the Hindu Right: 'science and technology, middle-class values which have long gone hand-in-hand with narratives of hard work and education, were thus to be the arenas within which competing claims to promote national pride in the face of this competition would be realized' (Nisbett 2009, 47). Nisbett argues that the rise of IT in India, the political ascendancy of the middle classes, and the discourse of IT as progress cannot be seen in isolation. They go some way towards constructing a career in IT as a natural aspiration for young educated Indians wanting to be part of a globally powerful knowledge society. The early 2000s saw a manifold rise in the number of private engineering colleges as a direct response to the IT boom, though the character of ownership remained the same. It signalled the extent to which existing managements were able to consolidate ownership, giving birth to education empires with a single management running several professional colleges, and sometimes even feeder schools.

In Tamil Nadu, the geography of aspiration becomes visible if one were to plot a map of the state based on the distribution and density of engineering colleges. Reflecting the urban reproduction of the professional class, Chennai and its suburbs[26] (northeastern Tamil Nadu) have the highest number of engineering colleges, followed by the industrial western districts. The southern districts, which are generally perceived as backward, volatile, and unruly given the caste disputes that have characterized the region since the 1990s (Gorringe 2012), have comparatively lower

---

[26] Northern Tamil Nadu that comprises the capital city of Chennai which has the most number of engineering colleges at 185. The district of Chennai has only ten colleges, but the urban centre spills into the neighbouring districts of Kancheepuram and Thiruvallur which have eighty-two and forty-three colleges respectively. Also, with high real estate prices in the city, it is probably more feasible to acquire the ten acres needed to start a college in these districts than the five required in the city. It is a common early-morning sight in the city to see bright orange college buses plying up and down, carting students between these three districts. The high number of colleges in and around Chennai also shows a clear preference among urbanites for professional studies such as engineering.

numbers of colleges. As of 2015, the western districts (Kongunadu) of Tamil Nadu, where I conducted research, have 179 colleges, most of which are concentrated in the heavily industrialized and urban Coimbatore district (77 engineering colleges). A relatively large number of colleges are also distributed in districts such as Namakkal (32), Salem (21), and Erode (17).

A direct correlation can be established between the number of engineering colleges in northern and western Tamil Nadu, and upward mobility of middle castes (categorized as 'backward class' groups) such as the Vanniyars and Gounders, who have been able to successfully diversify successfully from agriculture into textile, poultry, and trucking businesses (Chari 2004a, 2004b; De Neve 2006; Ravishankar 2015). This does not mean that every Gounder or Vanniyar is successful or that there are no poor Gounders and Vanniyars. However, these communities, on the whole, have acquired a certain visibility through their mobilization of, and engagement with various kinds of capital: from buying household goods, sports utility vehicles (SUVs), engaging in consumerism, securing membership of various clubs and philanthropic societies, and investing in higher education for their children, as mentioned in the previous chapter.

Investment in engineering colleges has been a chief means of securing this visibility, a trend partly enabled by increased reservation for Backward Class groups (OBC, MBC, and Denotified Groups). After the new system of reservations (twenty per cent of a total fifty per cent under reservation for Backward Classes would be set apart for Most Backward Class and Denotified Groups) came into effect in Tamil Nadu in 1989, the historian M. S. S. Pandian notes that Most Backward Groups such as the 'Vanniyars have increased their admission into professional courses six fold' (2000, 509).

Moreover, such groups have also politicized educational changes and are vocally opposed to mechanisms such as common entrance tests for admission in engineering and medical colleges as a form of inequality built into the system, a positive bias for urban elite students with access to coaching centres that help them crack 'aptitude-based' tests while 'shattering the dreams of rural youth' (*The Hindu*, 29 April 2016).[27] While

---

[27] The PMK has held this position over a period of time and opposed common entrance tests at all levels of professional courses. Also see *India.com* (2016), *The Hindu* (2017) and S. Vishwanathan (2005).

playing the politics of urban elite versus rural poor to its advantage, the PMK is opposed to reservation for Dalits on the basis that this group, once landless labourers, have taken advantage of employment and education sops given to them since independence and achieved parity, while Vanniyars have remained poor peasants. Such strategies are observed not only in Tamil Nadu, but across the country where the spread of private educational institutions is attributed to the rising aspirations of non-Brahmin elites, in a milieu in which higher education, government jobs, and professional occupations had once been dominated by the upper castes (Kaul 1993; Kamat 2011; Upadhya 2016a). At the same time, this has also resulted in intense feeling of competition with Dalits on the basis that reservation has helped them take strides in higher education and secure government employment, whereas 'backward caste' groups have remained 'backward', and still reliant on agriculture for their livelihoods.

The feeling of competition has also intensified the middle-class investment in education, with higher investment even for secondary education. For instance, a significant development in the 2000s has been the growth of 'super schools' (as they are popularly known) for classes eleven and twelve; private schools that are known for capturing the majority rank share in class twelve examinations. According to a report in *The Hindu*, a third of all medical and engineering places in the state are cornered by students from these ten to fifteen schools, which charged up to ₹2 lakhs (about $3000 US) in annual fees (Elangovan 2012) in the early 2010s. Their preponderance in the districts of Namakkal, Salem, Erode, and Dharmapuri speaks of the deep value given to professional education in the region, even though the schools also attract students from other areas. They are popular with middle-class parents despite high fees and the 'all-work-and-no-play' regimen, and newspapers carry stories of parents thronging these schools at admission time every year. These colleges promote the idea that the sole purpose of class twelve is to enter professional education: students are asked to state their preference between medicine and engineering at the time of admission and are trained accordingly. The implication is that this would, in turn, lead to successful careers in corporate spheres or in government. About a third of the interlocutors interviewed for this study had attended these schools, while others had attended 'branded' schools that had transformed from 'coaching centres' to large educational conglomerates, intended to help students crack national-level examinations such as the IIT–JEE (the joint entrance examination to gain admission into the Indian Institutes of Technology).

These schools are also a larger manifestation of the investment into the 'success' of the middle-class child, specifically the acquisition of economic, social, and cultural capital. As Nita Kumar (2011) points out, such an ideology of success is not produced solely at the site of the school, but primarily in the middle-class home, which treats the child as 'special' and invests specifically for his/her success. The politics of the family, which appropriate a host of techniques such as blackmailing, threats, and disciplining, ensures that the child cooperates in the scheme of things. While these have emerged in historically contingent ways (Kumar 2011), what stands out is that the home and the school appear interlocked in the production of the *difference* that delineates the middle-class child as specifically marked out for professional jobs in management, engineering, and medicine (also see Donner 2011; Sancho 2013).

However, the 2000s also mark the withdrawal of the elite from the Indian education system, with intensified migration to universities in the US, Europe, and Australia for higher education as well as teaching jobs, leaving many private colleges and universities with a lacuna in scholarship and expertise. Most existing teachers in the private colleges are products of the same or similar institutions, who regurgitate what they themselves have learned as students, with little exposure to innovations, global industry practices, or new trends. Many of them (especially women) are little older than the students they teach, their jobs in private colleges often a brief interlude between student life and marriage/more secure employment. Designated as 'management positions', they are paid about ₹20,000–25,000 per month ($270–$300 US—about a third of the government pay grade for university lecturers), and gain little in terms of employment benefits, making these less-than-enviable positions that do not attract real talent or scholarship. Moreover, apart from a few that are deemed to be universities, institutions come under one large regulatory body such as the Anna University in Tamil Nadu or the Visveswariah Technological University in Karnataka, and can boast of little innovation in terms of curriculum or learning models, having to follow the structure set by the university. Though some colleges have been granted autonomous status, many stakeholders say that it has led to the dilution of courses and streams rather than enabled them to gain in strength.

Though the numbers of private colleges exploded in early 2000s, there has been a steady decline in the number of colleges since 2016–2017 after fifty per cent of places started going empty, drawing attention to the bubble that characterized the rise and fall of private engineering colleges (Sivapriyan 2019). Though most critics have blamed this on the effects of retrenchment in the IT industry, it is also important to underline that there have been many other factors including the rampant corruption that has characterized the growth of colleges, their entrenchment in various other business interests that has not necessarily contributed to the enrichment of education, and the sheer growth of unemployment among engineering graduates, aspects which I explore in the next chapters.

Conclusions

By locating the multiple locations from which the proliferation of engineering colleges took shape in Tamil Nadu, I attempt to show the complex ways in which the private engineering college has circulated as a symbol of mobility, competence, development, philanthropy, and class making. The multiplicity of narratives, whether of family and caste histories or industrial lineages and political entrepreneurship attempt to sketch an 'economy of affects' (Richard and Rudnyckyj 2009) which shape aspiration. This 'economy of affects' moulds aspiration as something deep and personal, in connection with larger movements, even as it enables greater control for the market to encroach on notions of philanthropy. By delineating these narratives, tshe chapter shows the gradual shaping of higher education, especially in engineering colleges, as a 'politicized system', often 'militating against the inculcation of actual intellectual virtues' (Kapur and Mehta 2004, 27) because of the involvement of various social and political actors who bring their own visions of education to bear on the institutions they create. Thus, education has emerged as a service or good, even as institutions exploit government regulatory terrain and charge exorbitant sums of money, even as it helps individuals and organizations build their social stature as 'institutional' big men or institutions.

As we see in this chapter and the next, engineering colleges have become platforms for various kinds of organizations to cultivate a positive public image—as educationists, as patrons of the arts, as 'givers', and as men having positive relationships with technology. I argue that this format of being a *periya aal* reignites the debate of philanthropy versus self interest in the context of higher education, and the illustrations in this chapter and the next, open up a wide spectrum of practices that have been able to encroach on the notion of 'giving' in education, even as they look to making students subject to various ideological and moralistic positions in the guise of securing mobility for their families and communities. Though this is often characterized as the middle-class taking control over education, Kapur and Mehta (2004) argue that what it really marks is the massive disadvantage this has served the middle classes, whose expenditure on education has gone up several-fold since privatization. Moreover, apart from a few premier institutions, the education delivered by private institutions does not equip students with enough knowledge and skills that are valued in employment (see Chapter 3), and students end up spending more time and money accumulating more degrees, additional certifications, and trainings to signal their fitness for jobs.

I have made three main interconnected arguments in this chapter: one, how the idiom of 'giving' and 'service' characterizes the history of private education, whose strength (though now eroded), has enabled the success of many public figures who have been able to rework their *periya aal* personas to include those of successful educationists and modern technocrats. Two, how this has contributed to the building of a model of higher education in which colleges intersect with other business interests ('edupreneurship'), without necessarily enriching the field of education in terms of pedagogy and innovation. Though this has its roots in industrial families and their notions of philanthropy in the post-independence era, this model has grown in scope to include a whole spectrum of businesses, both legal and illegal. Third, I show how these developments have been important in cultivating an 'economy of affects' which has made it possible for various business families, caste associations, and religious organizations to enter, mediate, and influence higher education, playing an important role in how youth is shaped as a 'lived experience'.

2
'Edupreneurship'
Mapping Management Practices

In Tamil Nadu, many stories circulate that capture the ways in which business dovetails with education and philanthropy and constitute what I have referred to as 'edupreneurship' (entrepreneurship in education). In the varied nature of these narratives, the idea of education as a social good and edupreneurship as filling a vital gap in the delivery of this social good emerges a leitmotif. This chapter looks specifically at the proliferation of private engineering colleges in Tamil Nadu to examine the role played by 'institutional' big men in forging the culture of 'edupreneurship'.

A neologism coined by the business media, 'edupreneurship' is meant to describe various business, technical, and creative innovations in the field of education. Critically speaking, however, the word captures the nexus and interplay between higher education and spheres such as industry, business, politics, religion, real estate, construction, and even mining/quarrying that have come to characterize innovations in the education sector, the histories of which have been discussed earlier. This chapter focuses more on the contemporary and attempts to ethnographically capture the practices of edupreneurship after the 1980s, when private engineering colleges emerged as a viable enterprise, whose establishment was eagerly embraced as an act of patronage by industrialists, entrepreneurs, politicians, religious leaders, and other influential business people who could afford to mobilize the minimum requirements to start a college. Andrew Wyatt (2020) says this period was akin to liberalization for 'entrepreneurial politicians' in Tamil Nadu, who diversified into the education sector more than any other field of business—a trend that only accelerated with the IT boom. Not only did engineering colleges attract more financial backing, those with existing interests compounded their investment to benefit from the returns offered by the outsourcing model. Engineering colleges emerged as a strategic site of investment not only because it was lucrative, or combined business with ideas of philanthropy and social progress, but also because it could function as a 'cover' or 'feeder' for service-oriented businesses as well as other industries that function in the 'grey economy' with the connivance of those in power.

In other words, 'edupreneurship' captures the mobilization and circulation of different kinds of capital such as land, caste, class, and industry into and within the education sector. To understand these relationships better, I look at the profiles of various 'institutional' big men, their contributions, and their influence at various levels in the education sphere—from the microcosm of colleges and allied businesses to caste associations and party and electoral politics, along with a discussion on the implications of such ownership for higher education. These accounts are important as 'edupreneurship' in practice centres a key personality/personalities (*periya aal*) around whom a cult of loyalty and admiration is built. This *periya aal* could be the founder/chairperson of the institution itself, or another '*periya aal*' to whom the founder/chairperson of the institution wants to assert his or her relatedness.

Locating 'Big Men' in Gender and Community Formations

To become a 'big man' is a common aspiration in South India, the fulfilment of which requires a process of long investment, marked by leadership, acts of public altruism, and generosity through charitable institutions within his/her 'domain' (Mines 1996, 21–23; also see De Neve 2004; Alm 2010). Historically involved in activities such as temple-building, patronage of village/temple festivals, and client–patron relations within the *jajmani* system (Mines and Gourishankar 1990), big men have gained renewed importance in the contemporary context—in modern cities (Mines 1996) as well as in small towns and villages (Alm 2010; Jeffrey 2010; De Neve 2000). This is driven by the increased permeation of state-managed bureaucratic systems in the everyday—the negotiation of which needs leaders, patrons, and 'influence' (Price and Ruud 2010; De Neve 2000). In such a milieu, making use of contacts' 'influence' is not only the mobilization of social capital but also involves recourse to illegal means, bypassing what are seen as tedious bureaucratic norms and regulations, as a method of getting ahead (Jeffrey 2010; Alm 2010). Within the education sector, big men such as politicians are known to have the influence required to consolidate *and* disrupt long-standing formations of capital by enabling entry of less privileged groups into education. They are also known to critique elite notions of 'merit', lobby to bring about change, and create symbols of pride for communities. This is clearly manifested in 'edupreneurship': even as allegations of land grabbing, illegal construction (*The Times of India*, 4 December 2012)[1], substandard infrastructure (Venkatesan 2010), and exploitative fee structures (*The Times of India*, 24 January 2016) fuel a narrative against college owners, it has done little to tarnish the popularity of colleges, their managements, and the aspiration to enter the colleges. Founders and chairpersons have emerged as great leaders, admired among their constituents with whom they are seen to share a personal affiliation and rapport. However, debates rage whether their educational initiatives have been beneficial for local communities as a whole or have only enabled greater social dominance/control, electoral success, and profiteering by big men and their trusts (Jeffery, Jeffery, and Jeffrey 2006).

Though some colleges are signified as memorial spaces or dedicated to patron saints/deities and named accordingly, most personalities bank on the institutions they build to craft careers as 'big men' and name the institutions after themselves/their families. The nomenclature also reflects the unique subjective relationships that big men aspire to cultivate with their 'constituents'. For instance, well-known business families tend to assert their family names and identities because they view the institutions as part of a long history of patronage activities (Price 1989; Mines and Gourishankar 1990; Mines 1996). 'Big men' who intend to attract clientele from a specific community are known to name the college after the community or a symbol/deity revered by the community (such as 'Crescent' for a college run by an Islamic trust). The anthropological literature on 'institutional' big men (Mines and Gourishankar 1990; Mines 1996; Alm 2010) illuminates how these unique relationships are crafted,

---

[1] More than a hundred colleges in Chennai were found to be engaging in illegal construction, expanding their campuses on non-approved lands.

and how their acts of public altruism and patronage cloak self-interest and attempts at social dominance (Jeffery, Jeffery, and Jeffrey 2006).

To situate 'edupreneurship' within a conceptual framework, one must not only highlight the role of individual, the institutions he helms, and his leadership in mobilizing a community of 'constituents' (Mines 1996; Mines and Gourishankar 1990), but also delve into two main rubrics of anthropological scholarship: one, the literature on 'the gift' which locates 'giving' in a rich field of social and symbolic meanings. Two, the literature on (family) businesses, patronage, and its role in the substantialization of caste and the building of community solidarity. The two are closely connected because of the rich tradition of 'giving' among mercantile castes such as the Chettiaars, Parsis, Marwaris, and Patels, and the social and symbolic significance that come to be attached to the term, its functions, and implications.

Within South Asia, terms that build on the Marcel Mauss' framework of 'the gift' focus on concepts such as *'daan'* (donation) and *'dakshina'* (a gift donation or reward, especially given to Brahmins) and the moral universe they create within the brahminical life-worlds, overturning ideas such as gift reciprocity (Parry 1986). In the contemporary, the concept of *daan* (giving) itself has changed and is invoked in diverse contexts in India (Sundar 2013). However, many forms of gift-giving still have roots in the age-old systems of client–patronage relations. The literature on the *jajmani* system that structured the traditional village economy summarizes the exchanges that were mandated as certain castes provided services to the dominant landowning groups in exchange for grain or cash (Wiser 1988; Dirks 1976; Fuller 1989), creating a village community based on inter-caste dependence. Such patron–client relations based on giving and receiving, which may be interwoven with religious forms of giving or *daan*, have a range of functions and meanings. Haynes, who studied the involvement of wealthy Hindu and Jain merchants in a wide range of gifting activities in sixteenth-century Surat, argued that by 'channelling a portion of their economic capital into symbolic investments', donors established 'a series of relatively stable, multi-dimensional relationships' with important people, such as the Muslim rulers. But they also concentrated their efforts within their own communities: 'They built up their social reputations and economic credit largely through religious gifting' (Haynes 1987, 357).

During the late precolonial period, the rise of new warrior leaders and the lineages led to ritualized competition, investment in temples, and endowments of festivals on a scale never achieved before (Dirks 1987; Price 1989). In the colonial period, even as the British government took over religious patronage in a bid to gain local legitimacy as rulers through the manipulation of locally meaningful symbols of power, many industrialists came under the influence of Gandhi's nationalist movement and were motivated to strengthen the nationalist struggle. In addition, there were many instances in which big men invested collectively for the development of their individual communities.

These instances point to the role of philanthropy in the reproduction and regeneration of intra-caste solidarity as well as in the production of inter-caste differences or hierarchy, processes that gained momentum under colonialism. In many cases, building a solitary caste identity was a first step to organized efforts to claim upward social or ritual mobility—as I have earlier highlighted through the example of the Gounders in Kongunadu.

The embedding of ritual and economic exchanges within the caste system, according to several scholars, has weakened considerably since the colonial period, being replaced by relations of exploitation (Breman 1974). However, this does not actually mean that temple festivals and patronage have lost their significance in the organization of ritual exchanges and cementing social ties, but reflects the transition from purely 'casted' forms of giving to a more 'dynamic form' that includes an inter-caste 'village community' and kingly models (De Neve 2000, 513).

Geert De Neve (2000) points out that this is especially true in the industrial textile towns of Kongunadu, characterized by inter-caste labour relations, in which events such as temple festivals fulfil twin roles: for workers, fulfilling the need to belong to a community and a sense of equality through common worship of the goddess; for big men and wealthy industrialists, the securing of constituents and the chance to portray themselves as generous benefactors of the town. Thus, patron–client relationships between different castes still exist today in reconstituted forms (De Neve 2000; Cross 2014; Piliavsky 2014): they help to form a united and peaceful community and are also instrumental in acquiring social legitimacy for big men, who feel that their wealth must be redistributed in order to justify their dominance and control (De Neve 2000, 512). Forms of giving have also been institutionalized through the active participation of big men in organizations such as Rotary and Lions Clubs (also see Upadhya 1997; Ponniah 2017) and building of schools and colleges for the benefit of the locality. These contributions are generally valued highly, even though further contestations may also exist (Jeffery, Jeffery, and Jeffrey 2006) along with allegations of venality as mentioned earlier.

Though translated by Mattison Mines (1996) as 'institutional big *men*' because of the masculine persona associated with *periya aal*, the term does not quite preclude women. There are a few female *periya aal* in the posts of school principals, managers, and correspondents as elaborated by Mattison Mines (1996) in his work on big men in Tamil Nadu. Yet, in the realm of private professional education, there are disproportionately fewer female *periya aal*, though figures such as 'Amma' Amritanandamayi and the twin daughters of the late Jeppiaar (who was at the helm of fifteen institutions of the Jeppiaar group) have emerged as being very influential in contemporary times. The scarcity of women in roles of ownership and higher management points to the gendered control of capital, and the ways in which female edupreneurs are more likely to evolve from traditionally feminine roles as schoolteachers or 'mother figures', rather than through the route of family businesses and the nexus of business, industry, and politics. However, female relatives of the members of the management committee do figure in the everyday lives of the institutions, taking a personal interest in the running of the mess, the women's hostels, participating and organizing cultural events, and so on. However, these tend to get sidelined. Chandrakanthi Govindarajulu, the wife of G. R. Govindarajulu of the P.S.G (Periyaveedu Sama Naidu Govinda Naidu) is not mentioned in the PSG family memoirs *The Fifth Brother* (n.d.) despite being an educationist at the helm of many institutions for girls' and women's education as well as having a long philanthropic career through organizations such as the Red Cross (Govindarajulu 2018). Many women continue to have such careers—especially from communities such as the Nagarattar Chettiars who consider themselves of artistocratic lineages—they run women's hostels, schools, and are actively involved in the revival and patronage of various arts and crafts.[2] Though such actions by elite

---

[2] In contemporary Chennai, women from industrial families such as the Chettinad Group are a case in point. They are actively involved in running women's hostels, schools, and in the revival of arts such as Tanjore painting.

men and women are often characterized as being similar to Gandhian notions of *seva* or service (Skaria 2002), it is important to highlight that they also invoke a moral universe of piety, morality, and commitment to service that helps to reinforce the families' reputation and prestige, enabling a moral universe 'where commerce is possible' (Rudner 1994).

Thus, even though literature often highlights the role of men in business families and caste associations, women also play an important role in the substantialization of caste. In her study of Aggarwal women, Ujithra Ponniah (2017) argues that women facilitate the cohesion of caste and reproduction of family business through three strategies through their inter-strata 'social work', intra-strata socialization, and managing individuating aspirations of family members through marital choice, especially in communities and families that are associated with business. In her study of family businesses in Arni village in Tamil Nadu, Barbara Harris-White (2002) argues that women perform three important roles: reproduce and manage the capital-managing male labour; provide food as part of wages for labour; and get their daughters strategically married for business interests. In his study of Jain family business in Jaipur, Laidlaw (1995, 358) says that women's expressions of piety help build the family's 'credit' in a broad sense, and is important to build a family's reputation not just in society, but in the *bazaar* (market) as well (also see Bayly 1982).

In the case of higher education, I argue from my observations in CCT, that women family members play an important role in legitimizing and sanctioning female students presence in the college. They often perform a kind of 'emotional labour' (Hochschild 2003 [1983]) by walking around campus and 'advising' female students against wearing tight leggings or wasting their time 'hanging out'. They also take personal interest in projects such as the beautification of the garden and the fixing of hostel menus. In fact, it was highlighted to me with a lot of pride by the head warden that the family members ate food from the hostel kitchens whenever they were visiting, and often complimented the cooks or suggested changes.

Such involvement is not only visible in CCT, but also in another women's hostel run by a prominent family of industrialists in Chennai. The hostel is advertised as a 'home away from home' not just because of CCTV surveillance, twenty-four/seven security, and homely nutritious food, but also the personal interest taken by members of the Chettiaar family in ensuring that a temple ran within the hostel premises, with a Brahmin priest visiting to offer *puja* every morning. Women from the family also participate in cultural events, and judge contests like Best Hostel Resident among the incoming and outgoing students to recognize talents such as dancing and singing along with feminine virtues such as sari draping and grooming.

Such details are also highlighted in biographies of educationists such as Chandrakanthi Govindarajulu of the PSG family, whose piety in listening to Thyagaraja's compositions at 4.30 am is admired as much as the trees she planted in the PSG campus (Govindarajulu 2018). Thus, women's stature and roles in education are highlighted very differently from those of men's, the two seemingly occupying different planes: the former involved in feminine activities and projects becoming of '*samaja seva*' or social service, while men are highlighted through their cultivation of worldly power, influence, money, and politics.

Almost every college in the region is known to have women on their boards, mostly women relatives of the college chairperson. Their presence represents figurehead ownership to enjoy benefits such as tax breaks and other perks granted to women entrepreneurs by the state. These women are meant to be beneficiaries rather than benefactors and not likened to big men. However, some of the younger members appear to have more evolved roles, which include everyday administrative duties. This also stems from their own education: they have engineering degrees themselves and it is interesting to note how their 'empowerment' has taken place as a specific function of familial relations and education, rather than as individual achievement.

To sociologically situate a privatized institution one, therefore, needs to plot it in patterns of family ownership, its character as a family business, and its place within the larger bouquet of other family businesses, as well as other altruistic/patronage activities. This chapter surveys several engineering colleges in Tamil Nadu, intended to highlight distinct management practices in engineering colleges that are meant to reveal the circulation of the 'economy of affects' (Richard and Rudnyckyj 2009) in which campuses are enmeshed, from which we can trace student subjectivities. I start with a detailed reading of the biography of Professor G. R. Damodaran Naidu (GRD), one of the sons of PSG Rangaswamy

Naidu and the principal of the PSG Institutes of Technology from 1951 to 1970, a syndicate member of the Madras University from 1948 to 1981, and a member of parliament from 1981. I then move on to examine more contemporary figures such as the management of CCT and other college managers to draw a contemporary picture of 'edupreneurship' in Tamil Nadu.

Regional Patrons and Technocrats

G. R. Damodaran Naidu (GRD), who is considered a pioneer in the field of technical education, represents the reworking of the regional patron and 'institutional' big man into the figure of the modern technocrat valued today. He is an important figure whose life gives a sense of the politics in which private engineering colleges have become enmeshed. Not only was he an entrepreneur and the principal of PSG, but he was also a member of the Tamil Nadu Legislative Assembly for many years, and was later appointed as the vice chancellor of Madras University, spearheading many educational reforms in the state and eventually, the country. The coffee-table book on the PSG family legacy, *The Fifth Brother* (n.d., 53) introduces him thus:

> GRD truly believed that the purpose of a charity is to be able to spread the message of education, of a kind that would also provide a means of sustenance to one and all. He passionately believed that technical education had all the answers for empowerment and social change. GRD experimented with education as no one else had ever done. That much of what he did during his time has become the basis of change in technical education in the country is an endorsement of his vision and foresight. A great innovator at heart, GRD never missed a chance to design a product or process and even export it to establish PSG as a learning centre for technology education, but also one that practices what is being taught.

This vision for PSG was combined with firm views on policy and politics: 'We of this generation have an onerous responsibility of planning and providing a sound system of education and training. The country's growth has to be built on such foundations and not by wordy fireworks, or by organising the students for negative purposes', he wrote in 1969 (*The Fifth Brother* n.d., 54). These convictions of Prof. GRD were demonstrated during the anti-Hindi agitations of 1960s, when students all over Tamil Nadu took to the streets to protest the imposition of Hindi by the Centre.[3] 'In PSG, not a single day was lost in regular academics, and when the government itself declared holiday and asked students to go home, Prof. GRD made sure transportation was arranged for every student to go home' (*The Fifth Brother* n.d., 56).

However, this was not because he did not support the cause of Tamil: he was a pioneer of the system of translating English technical terms into Tamil with his journal *Kalaikathir*, which he began in 1948—an exercise now taken up regularly by the Tamil Nadu government (see Introduction in this volume). Neither did he ignore societal inequities: he is credited with starting numerous adult literacy campaigns as well as developmental activities for the poor, as 'social service' projects for PSG students to engage in. Moreover, he himself contested and won multiple elections to the Tamil Nadu Legislative Council as well as the Indian parliament. However, what we perceive is a deep distrust of grassroots politics, student mobilizations, and its characterization as being antithetical to student life and college, which must be invested in the project of 'discipline'. In GRD's vision, one could locate the beginnings of what we could call the crafting of the private college as a distinct 'other' to the democratic politics that characterized government-run institutions at that time. This is also highlighted in *The Fifth Brother* which points out that PSG had 'better discipline' (*The Fifth Brother*, 56) when compared to the two other engineering colleges in the region, both of which were government run, and the claim that PSG has the best record for least disruption of class and

---

[3] Efforts by the Indian government to make Hindi the sole official language after 1965 were not acceptable to many non-Hindi Indian states, which wanted the continued use of English. The Dravida Munnetra Kazhagam (DMK) led the opposition to Hindi. To allay their fears, Prime Minister Jawaharlal Nehru enacted the Official Languages Act in 1963 to ensure the continuing use of English beyond 1965. The text of the Act did not satisfy the DMK and increased their scepticism that his assurances might not be honoured by future administrations. As the day (26 January 1965) of switching over to Hindi as sole official language approached, the anti-Hindi movement gained momentum in Madras State with increased support from college students. The agitation spread all over Madras State, continued unabated for the next two months, and was marked by acts of violence. The Congress government of the Madras State called in paramilitary forces to quell the agitation.

'student disturbance' in all of Tamil Nadu (*The Fifth Brother*, 56). Such references to 'unruliness' and 'disturbance' are followed up by references to GRD's encouragement for sports and games—considered to give students confidence and build the rapport between students and teachers (*The Fifth Brother*, 56).

In GRD's vision of education, we not only see the allegiance to the Nehruvian model of the student body as something to be 'disciplined' through activities such as sports (Srivastava 1996; Jeffery, Jeffery, and Jeffrey 2006), but also an almost religious fervour for engineering and technical education. However, his identity was not limited to that of a teacher/educationist: like a true patron, he gave land to faculty members of PSG at 'nominal prices' cultivating two housing layouts to construct their own houses—Damu Nagar (named after himself) and Vidya Nagar (named after his son) (*The Fifth Brother*, 58). GRD also granted loans and machinery such as lathe machines for graduates to start their own industries: another big-man trait (De Neve 2004). Several PSG students are known to have begun pump-set businesses of their own, with GRD's patronage, contributing to greater mechanization of agriculture and wider use of pump-sets in the region (Chari 2004a); in fact, Coimbatore district had the largest number of pump-sets in the country right from the 1940s, a trend that accelerated by the 1960s, with increased electrification as well as newer pump-set businesses such as the Sharp Group of Companies, Mahindra Industries, CRI Pumps, Texmo, and Suguna Pumps (*The Fifth Brother*, 112).

In this way, the private model was shaped in PSG during GRD's time, not just through industry exposure and the inculcation of innovations such as the semester system, internal evaluation, and elective courses (*The Fifth Brother*, 56), but also through ideas of discipline which were strictly against student agitations and politics (*The Fifth Brother*, 55). Active political questions of the time were turned into issues of culture, and every institution today can probably boast of forums such as the Tamil Mandram, a trend that GRD first started in PSG. This is not only meant to be educational, but also an important aspect of showcasing the big man's patronage of arts and culture.

These changes in higher education set the stage for the changes post-1991, when they were widely embraced and became the new normal of student life, incorporated into a model of business that I term 'edupreneurship' that took the private model elaborated above and honed it further, as a response to the demands of the IT boom, globalization, and other social-economic changes. It must be remembered that PSG College of Technology is the only one among the first crop of colleges to have retained its identity as a private engineering college. The Madras Institute of Technology was taken over by Anna University in the 1970s. This is largely due to the efforts of GRD who modelled himself as an educationist rather than industrialist and kept the college running as his own personal enterprise. The fact that PSG is a benchmark against which industrialists-turned-educationists measured their own institutions was illustrated to me several times during fieldwork.

Education and Industrial Businesses

The links between industry and colleges are not relegated to the pages of history but find an everyday presence in the institutions as well. The Thiagarajar College of Technology in Madurai is owned and managed by a family belonging to a non-Brahmin mercantile caste group, and named after the founder. The engineering college, established in 1957, has a similar history to PSG Tech and grew out of a polytechnic institute that was meant to train labour for the textile industry. The founder of the polytechnic, Karumuttu Thiagarajan Chettiar was a contemporary of industrialists C. Rajam, Alagappa Chettiaar, and Rangaswamy Naidu (on whose lives I have elaborated in the previous chapter). He grew up in Ceylon and worked briefly as a journalist. However, he was forced to enter the family textile business when imports took a severe beating after the Great Depression, and traders had to turn producers. Thus, circumstances forced a capitalist out of an erstwhile journalist, whose report on the apathetic conditions of the Indian plantation workers caused great controversy in the 1910s.[4] His radical streak never quite left him, and he was at the forefront of the anti-Hindi agitations, even quitting the Congress party at the height of the struggle. He also cultivated the image

---

[4] A series of articles by him in 1913 in the paper *Morning Leader*, exposed the inhumane working and living conditions of Indian labour in the estates of Ceylon. These articles, and other representations, led to the appointment of the Marjoribanks–Tampi Marakkayar Commission by the Madras government to examine the travails of the Indian workers in Ceylon and Malaya.

of himself as a keen patron of the arts, particularly those related to the Tamil language and was even known by the sobriquet '*Kalaithanthai*' (Father of Arts).

The college in which I conducted fieldwork, the Chinna College of Technology (CCT) also has its roots in textile industrial capital and reflecting this: the college is one of the few in the region to offer a B.Tech in Textile/Apparel Technology. At several events held in the college, these features were discussed as the college's unique features with the degree in Textile Technology showcased as the college's flagship programme, and the family's industrial history[5] described as 'merit' in the subject. Given the family's 'experience' and 'influence' in the industry, they would not only be able to train students in the latest technology and best practices but also arrange successful in-campus recruitment through their contacts, it was said. At an Open Day event for prospective students that I attended, the college principal delivered an address to the gathered parents and youth that textile is to CCT what pump-sets are to PSG, and prospective candidates must consider enrolling in CCT if their chosen discipline is the textile technology programme, just as they would have considered PSG if their top choice was Mechanical Engineering. Moreover, he underlined that it is important to differentiate between such institutions and those run by corrupt politicians. The histories of PSG and CCT are illustrative of how old industrial capital is leveraged to start colleges by business families and is worked into a discourse of 'merit'—a term that has come to be politically charged and used to signify innate and organic talent by those opposed to caste-based reservation in the education sector.

Another piece of popular lore circulating at the college is the management's link to a key moment in the software history of India. Several interlocutors among the faculty, students, and management members told me that the group were also the builders and promotors of the first-ever IT park set up in Bangalore in 1984, a building that housed the first Indian offices of Texas Instruments, Cisco, Oracle, and Verifone. Various photographs of the event show the college chairman with the

---

[5] The old doyen of industry owned two dozen textile and sugar mills and the Madura Bank and started the concept of free education establishing half a dozen high schools and the Thiagarajar Colleges of Arts, Science, and Engineering between 1949 and 1956.

head of Texas Instruments, then Karnataka Chief Minister Ramakrishna Hegde, and others standing against the background of the first-ever satellite dish installed in India—a crucial moment for India's identity as an outsourcing hub.[6] Moreover, the family owns multiple businesses related to the global IT supply chains from software companies to agencies offering workforce training, HR consulting, and recruitment services. In addition to the engineering college and the polytechnic, the family also runs two schools and an institute of management.

Moreover, given their success as edupreneurs, the family's textile business, too, has been rebranded as a company specializing in the design and supply of uniforms for engineering colleges and corporate offices. As engineering colleges attempt to equip students to compete in a ruthless job market, the task of preparing students for global careers has become an important objective of higher education. Colleges are expected to 'craft professionals' out of students, not just through the inculcation of technical competence but to be socially adept as well. Along with the honing of communication and soft skills, colleges prescribe uniforms of Western formal wear such as a jacket and smart trousers at certain events to simulate the future workplace, and give students the embodied confidence to function in such settings (Hebbar and Kaur 2021). The unisex uniform, consisting of dark blazers and trousers, has become a ubiquitous but important marker of the aspirational culture inaugurated by the IT boom. Within such a context, students in engineering colleges have emerged as a ready market for the family's original 'thread to garment' enterprise—a business opportunity on which the management has been quick to capitalize.

The ownership of software IT parks, symbolic dominance over certain artefacts of technological advancement and professional success, industries and supply chains are important in communicating a symbolic mastery over the engineering arena that tends to be asserted as factors legitimizing control of education. These are not only highlighted through the family's involvement in national and international developments such as the inauguration of the IT park mentioned earlier, but also by engaging in local forms of patronage.

---

[6] An interesting titbit of information here is that the satellite dish was delivered by bullock cart to the Texas Instruments office. The contrast of the bullock against the aspirational architecture of the building was a striking one—a feature that continues to dominate much of the 'New India' almost forty years later.

Since the death of the founder in the early 2000s, CCT has been managed by male members of the founder's agnate kin, and continues to be enmeshed in family networks and client–patron relationships. As a mercantile caste that wielded considerable influence in overseas trade, the community prides itself as 'having stepped into the shoes of kings' in their patronage of temples (Thiagarajan 2004, 18).

Historically, the CCT family, too, is known to have extended endowments to temples all over Tamil Nadu and Ceylon. This idiom of patronage and kingship is worked into the everyday running of educational institutions. The college grounds occasionally host itinerant idols from temples—grand events at which hundreds of visitors come for *darshan*, and are fed as part of the family's patronage. At these times, the family played host to itinerant idol of Lord Balaji and the deity was installed on the sports ground, and the entire ground wore a festive look reminiscent of a village temple festival rather than a college. The college chairman performed various rites and rituals for the deity, and turns were also given to various representatives from the hostel community to make their offerings.

It was reported that 2000 visitors had come to campus that day, in addition to the resident student community, all of whom were fed at the expense of the family. These events, which mimic the 'village community' and kingly models (De Neve 2000; Price 1989) involve students, college employees, and their families, as well as local notables, becoming a means of developing and reaffirming a range of loyalties between parents and management, local administration, and members of the public. Such public events at the college are inscribed with meanings, and are important in showing how individuals and families are able to craft social dominance through their control over institutions. The involvement of the chairman and his family in recreating a village temple festival on the college grounds is one of the diverse practices of creating a community feeling within the college body, with the college chairman at the apex.

Politics and Authority

If industries and engineering colleges are closely linked in a variety of ways, the third part of that triumvirate is electoral politics. In fact, becoming a successful industrialist, starting institutions, cultivating the persona of a big man are all seen as planks to a potential political career. Politicians' presence in the education sphere is not limited to Tamil Nadu, but a phenomenon seen across India. The political scientist Rahul Verma (2020) argues that colleges are important nodes in the nexus between crime, politics, and bureaucracy, even as they signal patronage activities of the owner. It is said that between 2000 and 2015, six new colleges opened in India every day—a phenomenon which would have been impossible if the owners did not command power, influence, and possess the capacity to dole out bribes and kickbacks. Verma's (2020) examination of the power bases of politicians in Uttar Pradesh shows that over thirty per cent of politicians own a college, school, or both. One Lok Sabha member is said to run forty-five colleges. Colleges have become a necessary implement within the game of politics, as it signals not only patronage, but also an important source of money and muscle.

As discussed in the previous chapter, MGR's regime was responsible for a crop of edupreneurs from among the All India Anna Dravida Munnetra Kazhagam (AIADMK) elite in the 1980s. This crop of leaders had already contested and won elections by the time they founded the colleges. However, this trend has now reversed: coming to own an engineering college is no longer a 'kickback' received from holding public office. It is now considered that institutional ownership enables a political career because it provides a source of income that can fund the enormous cost of contesting elections. Ahead of the 2014 general elections in India, at least a dozen owners of engineering colleges joined the fray. Many of them declared assets worth over Rs 50 crore (about 60,32,000 USD).

Among recent MPs, former Deputy Speaker T. Thambidurai and his family are associated with at least five engineering colleges. SRM Group of Institutions founder T. R. Panchamuthu floated his own party, the Indiya Jananayaka Katchi (IJK), which sided with the Bharatiya Janata Party (BJP)-led National Democratic Alliance in the 2014 and is with the Congress-led United Progressive Alliance after the 2019 elections. A. C. Shanmugam (Dr Maruthur Gopalan Ramachandran University), one of the edupreneurs granted licences by MGR, returned to the election arena after three decades to launch his own party—the Pudhiya Needhi Katchi, which is an ally of the BJP-led Natioanl Democratic Alliance. Other notable owners include S. Jagatharakshankan (Bharath University) and Pongalur N. Palaniswamy (Kaliagnar Karunanidhi Institute of Technology) from the Dravida Munnetra Kazhagam (DMK).

In a survey conducted by the political scientist Andrew Wyatt (2020) on seventy members of the DMK elite who contested elections in 2011, at least twenty-one colleges were linked to senior members of the DMK in 2011. In analysing the various businesses in which politicians were involved, he says that the 'the larger scale activity of a subset of DMK politicians is concentrated in the education sector, and especially college education. These politicians tend to be older, have long political careers, often including a period as a state cabinet minister and/or a district secretary in the DMK' (Wyatt 2020, 7). Wyatt draws further attention to the connections of various DMK politicians to education, reinforcing the links between high party position and ownership of colleges. In considering their practices, one can also see how many of them use the educational trusts to cover ownership: 'Many of these colleges are said to be owned by educational trusts so the promoters can claim they do not own them, but the trusts are usually a device for distancing the colleges from the politicians that operate them' (*The Times of India* 2011b).

Even as certain politicians like S. Jagarakshakan and A. C. Shanmugam have returned to contest elections in changing political climates, others such as G. Vishwanathan of Vellore Institute of Technology (VIT) have been able to completely withdraw from electoral politics and have directed all their energies in crafting 'education empires'—chains of colleges that capitalize on brand value and have multiple campuses all over the country. Vishwanathan's VIT, for instance, has successfully marketed itself as an institution on an international scale, and has campuses in Madhya Pradesh and Andhra Pradesh in addition to Tamil Nadu.

Others such as Jeppiaar have more local appeal in the ways they have constructed a moral universe. A member of the Latin Catholic Mukkuvar community, a Most Backward Class (MBC) group, Jesadimai Pangiraj's (J-P-R which he used as an acronym) life and career is emblematic of the way in which an 'institutional' big man, specifically a 'entrepreneurial politician' is able to influence larger trends in education. Jeppiaar began his career as a police constable, and rose to prominence in the Anna Dravida Munnetra Kazhagam (ADMK), even earning the epithet 'MGR's right hand man'[7]. He served as chairman of the Madras Water Board, the Chief Whip of the Legislative Council as well as the Chairman of the Fisheries Board. Initially granted land and licences by MGR to start one engineering college, which Jeppiaar named after MGR's mother Sathyabama as a tribute, Jeppiaar grew in stature as an edupreneur. By the time he died in 2016, he was at the helm of an 'education empire' of fifteen colleges in and around Chennai city. He was also the president of a consortium of private engineering college owners, a pressure group meant to influence policy in the state in favour of self-financed colleges. An English daily once described him as one of the 'landlords' of Chennai (*The Times of India*, 18 April 2008), along with other owners of engineering colleges because of the extent of lands he owned in the city and surrounding areas.

Like his name, Jeppiaar's appearance too, was crafted to resemble MGR's: he wore dark glasses and dressed in white (the latter is a common aesthetic style among big men to signify their non-involvement in manual labour). When MGR died in 1987, Jeppiaar backed MGR's wife Janaki Ramachandran's faction in the AIADMK. Janaki became Chief Minister for a brief period, before the party was taken over after by J. Jayalalithaa. Thus sidelined, Jeppiaar began to focus on his education business. Apart from the engineering college named after MGR's mother, he started institutions named after his own mother, Panimalar, the Catholic saints, St Joseph and St Mary, as well as himself. Some of these institutions diversified to offer courses in management as well as the arts and sciences. In this way, he was able to successfully scale up from one college to establish fifteen institutions and secure various other business interests (software, water, salt, milk, steel, transport, cement, and a fishing harbour).

On their websites, institutions run by the Jeppiaar Trust carry a profile of Jeppiaar in which he is referred to as a 'self-made man' (Sathyabama Institute of Science and Technology website n.d) while simultaneously asserting his many connections to MGR and his commitment and service to the cause of education. Such a narrative plots Jeppiaar as a Robin Hood figure, much like the cinematic and political persona of his mentor

---

[7] In an interview to *The Times of India*, his daughter Maria said: 'Appa was like MGR's right hand. They shared a close relationship. In fact, when we were born, it was MGR who named us Anbu Nila and Thanga Nila. We were quite young when MGR passed away so we haven't interacted much with him but he taught us many things in life.' Reproduced in *The Newsminute* (2016).

MGR. Reflecting the popularity of this discourse, many online public forums allude to Jeppiaar with platitudes such as 'social reformer', 'education kingmaker', and 'philanthropist who democratised education' (Subramanian 2008). These sobriquets reference Jeppiaar's pioneering 'edupreneurship'—successfully piloting the model of the private college with gendered rules and the strict policing of the student body, while also batting for the disadvantaged and catering for their aspirations of becoming employable.

Jeppiaar campuses are modelled on the motto advocated by Jeppiaar that 'Entry (is) not important, exit is important' (Sathyabama Institute of Science and Technology website n.d.) meaning it is not important how one gains admission into engineering college, it is more crucial to exit with the skills required to be employable and successful. This was an especially powerful motto for those who entered college without existing cultural and social capital (usually about a third of the students enrolled in the college are the first-ever graduates in their families, another third enters through the management quota). As the president of a consortium of self-financed colleges, he used his influence with private sector employers to critique their hiring policies, which emphasized 'merit' (having consistent academic records, including in school examinations, exclusively English-medium schooling. and other forms of cosmopolitan cultural capital which lower caste/class and rural were unlikely to have). Instead, he asked them to focus only on performance in college and in the recruitment tests (*The Hindu* 18 June 2016). He believed that given the right kind of help in college, even disadvantaged students could do well in recruitment tests and rightfully earn their place in the private sector. It was a discourse of second chances, which struck a chord in many who faced the threat of unemployability because of the disadvantages presented by their backgrounds. As mentioned earlier, 'merit' is a politically charged category in post-Mandal India, a term often used by the elite to delegitimize access to education of the less privileged). Jeppiaar's plea was meant to break the vicious cycle of 'merit'-based education leading to 'merit'-based hiring, and level the playing field for everyone.

Jeppiaar is also known to have recruited extensively from his village and surrounding areas to fill non-teaching positions in the colleges (canteen staff, clerks, peons). The employees are provided board, lodging. and clothing, and are known to be unflinching in their loyalty to Jeppiaar, recreating the moral universe of the village: referring to Jeppiaar as *'thalai/ thalapathi'* (head/headman), reporting student violations of 'rules', and creating 'translocal' networks of village and caste communities within the campus (Subramanian 2008). In fact, Jeppiaar colleges are known for extremely strict enforcement of rules that do not allow men and women to speak to each other and 'professional' and 'desexualized' clothing for students. Students are known to be 'punished' with having to undertake clerical and menial jobs in the college if they are found disobeying the rules. In the 1990s, such an explicitly gendered and class-marked context of education emerged as a desirable option for many families who feared that their children's youthful behaviour in college, outside the realms of statused-adult society (Osella and Osella 1998; Nakassis 2016), would ruin their reputations. On the other hand, his colleges were known to mimic the moral universe of a village or small town, and therefore, perfectly suitable. In him, parents saw a reflection of their own moral authority, and a person of great 'influence' and 'power' capable of making their aspirations come true.

Jeppiaar's stature as a person with great influence was further cultivated through patronage of the arts, institution of awards (Jeppiaar Icon and Woman Icon Awards), organization of large public gatherings, favouring people from his town/constituency/religious sect for jobs and admissions to college, and (following another attribute of MGR[8]) feeding every visitor to his college. After his death in 2016, his 'education empire' has been taken over by his twin daughters. Maria and Regina Jeppiaar are involved in the everyday administration of the college, their spouses occupying important roles in the administrative board, and their photos and profiles are prominently displayed on billboards and on the websites of the institutions. They have taken the place of their father at the helm of award functions and other patronage activities, having successfully converted Jeppiaar's 'edupreneurship' into a family business.

Construction of Expertise

If industry represented expertise, especially in the post-Independence era, today the hallmark of expertise is consultancy. A college in Salem is

---

[8] MGR is known to have said 'Feeding is primary, education is secondary' (Kannan 2017).

run by a family of 'consultants', who boast of expertise in different fields through the various consulting firms they run. Their governing board has a father–son duo who run a consulting firm catering to civil engineering and construction, and another brother, whom I call DLN to protect their identity who has crafted a unique brand for himself as a provider of career guidance—another dimension of edupreneurship. The family was closely allied to CCT during its inception and then broke away to start their own college. Rumours floated among the student body that there was a huge rivalry afoot, but the families seem to maintain an affable relationship, at least formally.

I first came across DLN's name during fieldwork. While reading the morning papers, I noticed that every article on higher education ended with a comment by him. Intrigued by the man described as an 'education consultant', I looked him up online. His website said he had an office on a main thoroughfare in Salem. I called him to fix an appointment and took a bus to reach there, only to realize that his office shared the premises of a coaching centre I had visited earlier with my interlocutors, when they had taken a test as part of a promotional campaign for the Graduate Aptitude Test in Engineering (GATE) coaching classes (if students enrolled in the course, they would get a discount equivalent to the marks scored in the exam). While waiting for DLN to wrap up a consultation with a class twelve student in a small modular cabin set off from the waiting area, I glanced at the inspirational messages framed all around the room such a: 'If You Want to Succeed, Turn Your Life into it' and 'Do it now!'. In analysing similar posters in his study of youth and subjectivity in coastal Andhra Pradesh, Jamie Cross says that such 'installation(s) precisely encapsulate the notion of "self as enterprise", a project to be worked on and recrafted'; its managerial language capturing the new language of capitalism (Cross 2013, 125).

Initially, I thought that the messages were aimed at inspiring the many students who came to consult him on their education and career plans. However, the interview made it clear that DLN had 'made himself' in a similar entrepreneurial spirit. A gold medallist from Anna University and Indian Institute of Technology, Madras, in his younger days, DLN now owned and ran three educational institutions himself (a school and colleges for engineering and architecture), was on the board of several other colleges, organized coaching for competitive exams such as GATE and CAT in the institute adjacent to his office, and had a successful career as an education consultant. He offered 'counselling' for students who come to take his advice,[9] charging up to ₹500 per session, and travelled all over the state dispensing advice and lecturing in various colleges on higher education and career prospects. Thousands of students and parents solicit his advice. He claims to study trends, the 'scope' presented by various disciplines, and matches it to the 'aptitude' of those who seek his counsel. His advice is particularly valued in a context in which parents sometimes lack the social and cultural capital to help their children make informed decisions in the ruthlessly competitive world of capital, where certain professions and disciplines are privileged far more than others, but can also be rendered obsolete as quickly.

DLN spoke about starting out in the family business of cotton trade, before cashing in on the education boom in Tamil Nadu in the early 2000s.[10] He made his education consultancy official and started investing in education recognizing that it was on the verge of a boom. Though he summarily offered a critique of the rapid growth of private engineering education saying that there were at least 200 colleges more than necessary in Tamil Nadu (pointing to irregularities such as colleges without a single student enrolled), he also said that lives like his had been transformed by the education industry and the spirit of entrepreneurship he brought to it. He stated that he had watched a number of lives transformed by higher education around him: not just poverty-ridden students who have been able to get an education and embark on successful careers, but also politicians, real estate owners, and businessmen who have been able to benefit from the entrepreneurial wave in education following the IT boom. 'Even "a peon" at the Directorate of Technical Education in Chennai has started his own an engineering college', he said. The statement reveals the derision of the 'educated elite' towards the new patterns of ownership that go beyond the traditional hold of privileged class/caste over the mechanisms of education that privatization has enabled, and the discourse of 'corruption' through which these forms of ownership are characterized.

---

[9] His website claims that he has counselled more than fifteen lakh (1,500,000) students.
[10] DLN, educationist and consultant, in discussion with the author, August 2014.

However, DLN's own life also reveals how notions of entrepreneurship have enabled him to create an image for himself as a renowned education consultant, rather than just a successful trader or industrialist in his constituency who invested in education. The spirit of entrepreneurship is important to flag, especially as it embeds itself within traditional occupational forms such as trade and industry. These traits of entrepreneurship mark college owners as distinct (also see Kaul 1993) and show how individuals or groups that already command certain power in the existing social system/class structure, either through achievement or ascription, often get involved in education playing the role of 'education mediators' (Dale 1982; Lynch 1990).[11]

Edupreneurship as Capitalization of Resources

Many critics of the ways in which education has been privatized in India have suggested that the only reason private institutions are able to have a financially sustainable model is through the various government subsidies offered to colleges. Indeed, several governments in South India (Tamil Nadu, Telangana, Andhra Pradesh) offer free education to members of disadvantaged groups. Thus, even though privatization is often characterized as the withdrawal of the state from a particular domain, the truth is that a considerable amount of government spending on higher education goes into supporting private colleges (also see Kapur and Mehta 2004). In fact, colleges are known to recruit aggressively among students from disadvantaged groups such as Scheduled Castes and Scheduled Tribes in order to collect the monies allotted for them. These colleges also offer incentives to students already enrolled in the college (Dayashankar 2019) to bring in students from other states as well as from their respective neighbourhoods and communities. With more than half the places going empty and some colleges struggling to register even 100 enrolments, colleges have hired agencies to recruit students from South Asian and African nations. Moreover, laptops, smartphones, free bus transport, and even gold coins/biscuits are offered as incentives

---

[11] Katherine Lynch (1990) examines the role of the Catholic Church as an 'education mediator' in Ireland.

to students to enrol (Dayashankar 2019). On the other hand, successful colleges in which admissions are sought after can command hefty 'management fees' and cultivate a brand value which is centred on providing infrastructure such as labs equipped with high-end widescreen computers, branded learning software, food courts, swimming pools, sports complexes, and air-conditioned hostel rooms. These are some of the key ways in which the 'education market' has catered to the spending power of the middle classes.

Even as these innovations attract more enrolments, the business of running engineering colleges meshes with other interests. One of the most striking examples I was told, were the links between the sand mafia and private colleges: the acquisition of land for a college provided an easy cover for the illegal mining of sand, without having to procure the requisite licences or environmental clearances. Once the site was excavated, an engineering college was built. The institute also provided a legitimate method of converting 'black money' from mining to 'white', after which partners in the mining business would become colleagues on the college management committee.

Yet there are forces of altruism at work, ploughing money back from mining into education: O. Arumugasamy, who has the reputation of being the lynchpin of the sand mining business in Tamil Nadu with a daily revenue running into millions, claims to have relinquished all profits from his legitimate businesses (paper mills, a cinema theatre, a biotechnology unit, and consulting services for mining businesses) to pour into higher education (Bhagat 2012).[12] Initially offering scholarships to members of his caste group in four districts, his Shree Vijayalakshmi Trust now makes direct cash transfers to any student who has secured admission in a college in Tamil Nadu, totalling to about ₹300 crore (about $37 million US). Around the time class twelve results are declared, his trust organizes a veritable fair to dispense the funds. While some students get full scholarships, others receive smaller amounts to supplement the payment of management and tuition fees. His motto is that no one should leave his premises empty handed, Such efforts at redistribution of wealth through

---

[12] O. Arumugasamy is the chairman of the Senthil Group of Companies, which runs a film theatre, owns paper and packaging companies, a film distributorship, a construction firm, and a company that exports papain (an enzyme from papaya used to tenderize meat). In his own words, he is nothing more than a consultant to those in the mining business (Bhagat 2012).

the idiom of 'giving' to higher education are important in justifying the enormous wealth cultivated (often through illegal means) and creating a sense that big men have the welfare of the community in mind. Such models bolster private models of college ownership as the money is 'given' to students creating loyal and grateful bases at many levels. However, not all business interests that mesh with engineering colleges are as sinister. Edupreneurship, in essence, has been the shaping of education according to the aspirations and anxieties of the various stakeholders in ways that can intersect with existing holdings of the management, their reputations, and plans.

Many edupreneurs, especially those whose 'constituents' lie in particular communities or have the following of certain 'sects', use college grounds to build places of worship. This is not the case of an occasional religious event as the case of the CCT temple festival or a small chapel as is the case of missionary-run institutions, but huge projects that are newly constructed to attract tourists. For instance, one campus in Salem boasts of a temple of 1008 lingams,[13] built by the owner of a college, comprising an entire hillock adjoining the campus. Each of the 1008 lingams forms a continuous stream all the way up the hill. At its apex, the 1008th lingam is installed and represents the main deity, Sundara Arunachaleshwarar with his consort Uma. Visitors said that the shrine, built in 2010, was intended as an act of penance by the college chairman for removing a pre-existing shrine from that site to build the college. I could not find any written source to corroborate the story, but the chairman enjoyed the status of having restored a simple shrine to greater grandeur. The project is also a beneficiary of the 'Incredible India'[14] campaign to market specific tourist sites in India, and snapshots of the temple appear in promotional campaigns as one of the key tourist attractions of Tamil Nadu. Many interlocutors said the site has been instrumental in putting Salem on the tourist map. The college was founded in the 1980s in the name of a particular seer, whose followers were mainly from the Asriyar Vaishya community. The current chancellor, an academic, who was chosen by the board to be the founder's successor, is also the president of a globally

---

[13] The phallic symbol associated with the Hindu deity Sivan/Shiva.
[14] Stylized as Incrdible !ndia', it is an international tourism campaign launched by thr government of India in 2002 to promote tourism in India.

linked association of the community showing how a successful career as an educationist can indeed lead to influence and dominance over an entire community, even transnationally.

In addition to such displays of religiosity, one should also consider the multiple 'displays of science' that characterize these spaces: solar parks, geodesic domes, futuristic models of vehicles, and model robots are often displayed on campuses. Some of these such as solar parks are not only meant to signpost the transition to cleaner and greener fuels, but also meant to capitalize on the tremendous subsidies and tax breaks offered by the government for tapping into solar energy (Sushma 2012). What they enable is also a 'spectacle of engineering'—artefacts that enable an imagination of engineering education as a commodity, meant to arouse feelings of awe, aspiration, and entrepreneurial spirit. Many are student projects: designed and put together from personal funds or funds from the college, which may or may not have utilitarian value for the college but are intended to inculcate an experimental and entrepreneurial spirit in students. Colleges usually promote such projects for publicity: they find mention in city pages of newspapers, especially in smaller cities like Salem on quiet news days. Students also take these models to competitions in other colleges, and they become important in improving the brand image of the college as an institution that facilitates research and innovation by encouraging students to take up projects outside the curricula. This trend which has always existed informally in private colleges (see PSG case study in previous chapter and in GRD's profile), though, has now acquired a new idiom of expression: college managements brand themselves as 'venture capitalists' and 'angel investors' incubating nascent businesses of students.

With the popularity of shows such as Shark Tank on national and international television, every college now hosts its own version, meant to train students to design their own business models, pitch it to investors etc. Students that succeed get access to office space, infrastructure, and mentorship by seasoned entrepreneurs from the management team to start their businesses, enable production, and capitalize on profits. Such a formalization of relationships between the college managements and students is a reworking of patronage, in newer forms and idioms, creating new paths for the circulation of capital. In such reconstitutions of relationships, 'students' are not just 'students', they are 'constituents' to be cultivated, and probably even partners with the potential for business.

Thus, contrary to popular conceptions of private colleges as only 'service providers' who view students and their parents as 'customers', I would like to suggest that managements' relationships with the communities they inhabit is crafted in a variety of ways: even as they continue to retain continuities with older forms of patronage, kingly models, and prestige-enhancing practices of business families, they also acknowledge newer roles and forms of doing business in which engineering colleges form one among a variety of business interests of the management. Given how colleges are but one unit in a string of enterprises that rely on local forms of labour and loyalty, a college manager today might even be an employer tomorrow. These newer meanings and forms of relationships are intended to heighten the respect and admiration that colleges managements have in a community, and therefore rework the relationship between the state, self, and the market in newer ways within what Richard and Rudnyckyj (2009) would call an 'economy of affects'.

As they point out through two different neoliberal climes, 'affect' produced through acts such as embracing and copious crying play an important role in neoliberal transformations. In contemporary India too, 'edupreneurs' such as 'Amma' Amritanandamayi have been able to wield vast spheres of influence including in arenas such as higher education through similar acts of hugging, in the context of a state receding from various domains of public service. They inspire donations including large tranches from countries such as the US (where they are registered as 'churches' that are tax exempt) as well as inspire people to work for free, able to effectively bring about desirable worker subjectivities by building the notion that they engage in crucial acts of public 'service' (Halpern 2013).

Therefore, to recast the economy of education, particularly privatized higher education, as an 'economy of affects' presents not only a productive lens for historical events as these two chapters represent. It is also an important framework to evaluate the wider implications of private higher education in crafting and sustaining a distinct youth culture that intends to produce desirable 'employable' and 'global' subjectivities, but at the same time, prioritizes local and cultural norms within which the prestige of a big man is showcased. Within the domain of private engineering colleges, notions of caste, class, and gender are crucial in deciding and shaping potential careers based on what constitutes 'respectable' labour as well as a desirable 'employable' subjectivity. It is within this 'economy of affects' that both the effectiveness of education as well as the various subjectivities in education need to be evaluated.

Subjectivities in Higher Education

Various studies have shown that individuals who head institutions are sometimes larger than life, believed to be 'the sole repository of the virtues and vices of the institutions' (Sudhir Kakkar cited in Mines 1996, 45). Institutions, too, are viewed as the repositories of the virtues and vices of individuals, families, and communities. They can make or mar a family's reputation, enhance prestige, and create a moral universe in which communities are constructed not only along the lines of caste, but also in terms of employment, benefit, and expertise. In other words, institutions create the larger material and moral world in which individuals assert their control and legitimacy over education and industry.

During my stay in Salem, such intersubjectivity between individuals and institutions was beautifully illustrated when a local tabloid carried a story reporting the alleged scandalous behaviour of the CCT hostel warden in his private life. The story, published with a bust-length portrait of the warden and a picture of the college in the backdrop, raised questions about how CCT can have a good atmosphere for study when it was governed by such individuals. This publication became the reason for tighter regulations concerning students' behaviour and conduct. Some students, when shown the tabloid article, suggested that such bad press was being generated by a rival college, with the malicious intent of harming CCT's reputation.

As a result of this kind of scrutiny on individuals and institutions alike, respectability is the dominant ordering principle of engineering college campuses across Tamil Nadu—the majority of which practise gender segregation on campuses. Teachers and members of the non-teaching staff are tasked with the labour of policing cross-sex interactions. Some institutions, such as the Jeppiaar institutions, forbid all forms of cross-sex interaction and enforce strict dress codes.

The ordering of institutions with big men from a certain denomination at the helm, also creates a certain kind of subjective clientele—of gender and caste/community, specifically. To elaborate on this, I present here a narrative by an interlocutor whom I call Smitha, who was enrolled in the Information Technology (IT) course in CCT. Smitha's younger brother Sethu had joined an engineering college in the middle of my fieldwork in 2014: her father paid ₹5 lakhs (about $6100 US), just so he could gain admission to a college that is owned by one of the top industrialists in Kongunadu. She narrated thus:

> When my father worked in a factory, I was not able to pay the monthly fee for school and would be made to sit out of the class. There was never anything to eat at home … My mother would make a thin *kanji* (rice-based gruel) in the morning and at night. Everything changed when my father realized that we should get back to the community activity of textile—that is what suits our community the most. But we didn't have money to buy equipment. Then the bank started a scheme for women entrepreneurs and my mother could take a loan, but she had to go for training class. We bought a spinning machine. We put it in the main hall of our house. There was no place to move, and my parents worked their machines to spin yarn all day. We then had money to buy another and then another … Employ more, build a workshop outside our house. We changed schools—I went to one of the most elite schools in Tamil Nadu, with extremely high fees … from the government school in which I could not even pay the fees! That is how much my life changed! After this much suffering and coming up in life, my father felt that his son should go to a college that is well known in our community, even though he had gotten a counselling seat (government quota seat) in another college. That is our pride—that's how far we have come!

Smitha's narrative points to the gendered decision-making by parents in response to caste-specific ownership of the colleges, and the gendered ways in which mobility is highlighted. In Smitha's father's case, investment into specific colleges for his son and daughter was intended as the presentation of self to the larger community—his decisions reflect the patriarchal control of his daughter, as well as the advantages he could give his son. Smitha's father felt the need to invest, not just in any engineering education, but in an engineering college that was managed by his weaving-community members in order to increase his own status. He could have chosen to take a college seat that came through the centralized system of the government, or another college that demanded less capitation fees (he had paid only ₹3 lakhs (about $3650 US) for Smitha's seat for a college close to their home), but instead he had chosen to invest in a college for his son that resonated with him as a college that was run by the 'pillars' of his own community, a family who also owned a sugar industry that traded on the national stock exchange. He saw that as a sign of achievement, success, and pride in his own toil; for him, the college was not just a monetary investment but also an emotional one, signalling his status in the community. From the case of Smitha's father, one can also analyse that private engineering education panders to what has been described as a common 'householder' subjectivity of men in South India—in ensuring that children get a good education, while seeking out a patron (De Neve 2004; Osella and Osella 2006).

Moreover, in this narrative, the lack of resentment against the personal gain for the owners/management for collecting capitation fees for college admission, reveals its normalization. Moreover, it remains concealed within a larger universe of community and individual pride in toil, and in upward social mobility resulting in status. College managements can exploit such notions to increase their own political and socio-economic power veiled as altruistic projects of imparting higher education. Therefore, the starting and running of engineering colleges by business families is not just the concrete extension of family-as-reproductive-unit into the market economy as Harriss-White (2001) argues, but the extension of networks of caste, patronage, and community into processes of social mobility and transformation through the education market.

Interviews conducted with interlocutors in the college also reflect investment in certain kinds of caste cultures. Meghana, for example, a Tulu-speaking Brahmin from coastal Karnataka whose family had migrated to Tamil Nadu to set up a hotel in Rajapalayam district many decades earlier, explained that her choice of college was purely based on the availability of a vegetarian mess. She said that her parents and grandmother had prioritized brahminical adherence to vegetarianism over a preference for an engineering stream and had been prepared to pay a management fee accordingly. Interestingly, CCT was one of few colleges to have a mess that serves only vegetarian food—again, a point of great moral pride for the management, with a chapter dedicated to vegetarianism in the founder's biography.

Others, too, indicated such an investment in a caste and gendered logic: daughters were to be held close in a hostel and college that reflected the protectionist logic of the home. Sons had to take their place in the world, hopefully with enhanced status. This, as I will elaborate later, led to certain consequences for gender and sexuality, and the way college culture is shaped. This was made possible by colleges because, as Lynch (1990, 12) suggests, when left to the mediators (that is to say, in return for the investment into education and its silent promotion of capitalist culture), they are allowed to propagate their own ideas and ideologies in the college and socialize young people as they see fit, as long as it adheres broadly to the propagation of technical education. This highlights the role that emotion or sentiment plays in such decisions—of enrolling in colleges, of starting new ones—and the importance of recasting the economics of engineering colleges with ones of patronage, sentiment, and affect.

Conclusions

The reason for highlighting various modes of doing 'business' through patronage and public altruism is not to exoticize certain practices as the 'Indian' or 'Tamil' way of education, mediated by personality cults, but to show the ways in which private education has crafted unique subjectivities through its mobilization of an 'economy of affects' (Richard and Rudnyckyj 2009). The numerous case studies discussed here show a plethora of practices in which higher education has built on existing formations of religion, caste, sect, industry, employment, and entrepreneurship. They have extended and reworked existing relationships to create a moral universe of giving, taking, community-making, working, and studying, and make possible the circulation of certain ideologies and forging of subjectivities that renew the contexts in which youth is experienced and lived by students. Thus, we see in GRD Naidu's conceptualization of higher education, a model of subjectivity that encourages the Nehruvian idealism of nation-building, a belief in technology, utilitarian value, and 'apolitical' engagement with development. For religious seers and caste associations, institutions are meant not only to strengthen community ties and initiate new followers to the philosophy/faith, but also enable a certain visibility for the group by engaging in activities that are considered to be for the 'social good' such as building schools, professional colleges, and hospitals. Even though many allegations of venality and controversy surround these figures, debating whether they should be exempt from taxes, or receive money from abroad, it is also well acknowledged that they are a crucial mechanism through which education, especially professional higher education, is accessed in India. In neoliberal India, many of these tropes of religion and sect have been reworked into an idiom that is in tandem with globalization, entrepreneurship, and the 'spirit of capitalism'.

This is also seen in the transition from an inflection on industry to individual entrepreneurship. In the case of CCT and other contemporary illustrations, we see the quest for legitimacy for control over education asserted through industrial expertise, niche knowledges, control over the global supply chains, adaptation to newer technologies, in addition to families' engagement in local economies of patronage and morality. Though these are important in producing a globalized workforce for the market, it must be underlined that allegiance to dogmatic/local ideologies have produced crucial gaps and deepened existing inequalities, even indirectly contributing to what can be called the engineering college bubble, creating a mania around engineering education while tied in a race to appear to be doing 'social good'. This certainly begs the question of whether the boom in engineering colleges has been truly beneficial as a 'social good', or has it only strengthened existing inequalities and worsening unemployment, while enabling hundreds of 'big men' to further their careers, business interests, and deepen their nexuses?

I explore some of these aspects in forthcoming chapters, looking at concerns around 'employability' as well as the production of gendered structures/processes of self-making to understand young people's own sense of their education, career prospects, and personal transformations to evaluate the meanings of education enmeshed in an 'economy of affects'. The end of this chapter marks a move away from a focus on 'big men' and 'edupreneurship' to explore aspects relating to the everyday lives of students, from unemployment, gendered constraint, as well as personal subjectivities. They show the extent to which the 'moral economy' within which 'affects' circulate is an important underlying structure through which young people make sense of their everyday lives, shaping choices as well as manufacturing notions of reach and restraint.

3
Becoming Professional
Dilemmas in Emerging 'Employable'

One morning in mid-August when I went to the Mechanical class, Catherine was bursting with some news. Even as the teacher started her lecture, Catherine started whispering to me under her breath. 'Have you heard?', she asked. When I replied that I had no idea what she was talking about, she said, 'The whole college is talking about it. How do you not know?'.

'What happened exactly?', I asked her again, really clueless why she was so excited.

'You know Sugam?'.

'Yes, of course'. Sugam Khatri's name was something I heard day and night since Catherine became a close interlocutor. He was a senior who Catherine was infatuated with.

'He got placed. Campus Placement in LG! They interviewed eighty people, and they finalized only one candidate: Sugam!'.

'Oh wow!'.

'It's because he is good at studies and has a well-rounded personality! Everybody is talking about him. In the canteen, in class, all our teachers are also praising his name!', Catherine said, brimming with pride.

'He has become a hero in our college!', she continued. 'One out of eighty, Mechanical student, getting placed in a core company! You know how difficult it is to get placed like that!'.

Catherine was not exaggerating when she said fourth-year student Sugam was the celebrated person of the hour. His name was repeated many times over in Chinna College of Technology (CCT) as the person who overcame all odds to get placed. His achievement was the stuff that engineering aspirations were made of. In engineering colleges across

Gender, Caste, and Class in South India's Technical Institutions. Nandini Hebbar N., Oxford University Press.
© Nandini Hebbar N. 2024. DOI: 10.1093/9780198914488.003.0004

Tamil Nadu, students have their hopes and dreams pinned to the on-campus placement drives, when companies visit to pick potential employees among the outgoing batch. Sometimes, these drives even begin late in the third year so companies can have an early picking. Indeed, a college's reputation is built on placement success and prospective students and parents consider placement records very keenly. CCT, for instance, advertised '100 per cent Placement' on almost all its promotional material, on billboards, and in promotional stalls set up in public spaces such as railway stations and education fairs, both in India and abroad. Many said that they took this promise of '100 per cent Placement' very seriously, and were certain that their investment in the form of high tuition and management fees would reap returns in the form of well-paying jobs. In fact, across campuses, there is visibility accorded to successful candidates: their photographs are prominently displayed on flex boards along with their names, department, and details of placement. Such flex boards of various colleges are displayed across the city—outside the college gates, at the bus stand, just outside the railway station. They are not just advertisements; they are the very artefacts of success, representing a portal through which young people aspire to cross, but which also presents a world very different from the sheltered existence of home and college, drawing apprehensions about gender and class. Aspects of college life such as 'employability training', which this chapter focuses on, are important in showing the various tensions that characterize concerns about employment. They represent a fraught zone where locally accepted norms of respectability and prestige are refashioned to desirable 'employable' subjectivities, drawing a range of reactions ranging from gendered moral critiques of 'IT culture' to engagement with consumeristic 'professional styles'.

In this chapter, I discuss how the employability crisis and the ensuing pressure on education to make students 'employable' in the global knowledge economy has shaped engineering education, its subjective experience, as well as its representations and portrayals. Drawing upon the idea that education is a 'contradictory resource' (Levinson and Holland 1996; Willis 1977) that equips each student differently, sometimes with contrasting outcomes, I explore modules and lived experiences of employability training in CCT to understand how students 'creatively occupy'

(Levinson and Holland 1996, 14) the space of employability training. Everyday life processes, practices, and performances in the college illustrate the highly unstable ways in which 'employability' is constructed in colleges, because many attributes constituting 'employability' emerge in opposition to other frames of reference such as respectability and good standing at the local level. This paradox at the heart of employability training has accentuated existing forms of inequality such as caste, gender, and class.

The Placement Race

Though institutions such as CCT aim to recreate a training programme in line with what prospective employers such as global IT companies expect, they also have their own position, situated between the local and the global, in models of 'edupreneurship', the identities of the 'big men' running the colleges, and their respective social networks. This position is critical in producing 'disjunctures' between what employability training aims to inculcate and what is imbibed by the students. In the words of Arjun Appadurai (1996), mapping 'disjunctures' is crucial to show 'how colonial processes underwrite contemporary politics, how history and genealogy inflect each other, and of how global facts take local form' (1996, 18). Apart from their individual caste and class positions, students' geographical locations, circumstantial predicaments, subjective experiences, and the local context of the college influenced how they respond to employability training.

The table below of the break-down of the student body at CCT from the academic year 2013–2014 gives a sense of the backgrounds of the students. Across the three batches at CCT (total 2806 students), there was a total of 976 First Graduates (first to graduate in their families) enrolled. Their department-wise distribution and further distribution by category and gender is given in Tables 3.1 and 3.2. The bulk of First Graduates come from Backward Class (BC) and Most Backward Class (MBC) groups. These are also groups that invest the most in higher education as indicated by the strength of their numbers in the Management Quota (Table 3.3).

Table 3.1 First Graduate by Stream

Engineering Stream	Percentage of First Graduates
Civil	30%
Computer Science	37%
Electronics and Communication	30%
Electric and Electronics	39%
Fashion Technology	29%
Information Technology	37%
Mechanical	37%

Source: CCT Admission Data 2013–14 shared with author by college management.

Table 3.2 First Graduate by Category and Gender

	Male	Female	Total
Other Caste/General	15	10	25
(Other) Backward Class	256	230	486
Most Backward Class	160	112	272
Scheduled Caste	110	77	187
Scheduled Tribe	4	2	6
Total	545	431	976

Source: CCT Admission Data 2013–14 shared with author by college management.

Thus, even if most students at CCT have the financial backing to access higher education,[1] they still face some historic disadvantages and do not necessarily have the contacts or possess the skills to find jobs easily, especially in multinational companies. If a student seeks a job but does not get one during campus recruitment, the next few months would have to

---

[1] Parents of students I met at CCT comprised a very broad range: groundnut and sago cultivators, district-level politicians, textile business owners/employees, construction business owners/employees, casual labourers, grocers, blue-collar government employees, teachers, college professors, and even a wealthy mall owner. There has been much debate about what kind of background favours success in the IT sector. Krishna and Brihmadesam (2006) state that parents' educational background is the most important criteria determining success of a software engineer. In a study they did, seventy-five per cent of software engineers had fathers who were graduates.

Table 3.3 Management Quota by Category and Gender

	Male	Female	Total
Other Caste/General	76	45	121
(Other) Backward Class	367	256	623
Most Backward Class	183	71	254
Scheduled Caste	12	7	19
Scheduled Tribe	0	0	0
Total	638	379	1017

Source: CCT Admission Data 2013–14 shared with author by college management.

be spent trying to secure internships, or looking for jobs in the informal sector (coaching centres, taking freelance projects conceptualizing undergraduate students' assignments/projects, writing graduate theses). Desperation drives some graduates to 'agents' or brokers who promise 'lateral entry' into big tech firms (usually not open to candidates who have freshly graduated) in exchange for a 'cut' (commission percentage) from the salary package. The situation was made more poignant by the fact that higher education in engineering is used by upwardly mobile families to scale the social ladder, and many sell an acre or two of agricultural land to afford a place at engineering college. Moreover, a third of the students in these colleges are also beneficiaries of government-sponsored schemes such as the First-Graduate scheme and other subsidies offered to the students of disadvantaged groups to bring them into education.

While young women's conversion from higher education to employment remains dismally low due to various kinds of family pressure (see next chapter in this volume), the real crisis in unemployment in India is seen as a crisis in masculinity (Jeffrey 2010; Jeffrey, Jeffery, and Jeffery 2008). The breadwinner ideology posits men as primary breadwinners and young men not having a source of income are a liability to the whole family. Young men are therefore deeply pressured to find employment. If unemployed, educated young men rarely return to traditional occupations such as agriculture because of the associated lack of status. According to data released by the Centre for Monitoring Indian Economy (CMIE) in July 2022, the unemployment rate in India is 7.3 per cent (Benu and Satish Kumar, 2022). However, the unemployment rate among graduates is 17.8 per cent. Caste and class undoubtedly complicated this further.

A 2017 study concluded that of 8 lakh (8,00,000) engineers graduating every year, sixty per cent remain jobless (*The Economic Times*, 18 March 2017). 'Employability' rates remain as low as fifty-two per cent among engineering graduates (Skill India Report 2018, 21). In her in-depth study of IT companies and policies, Carol Upadhya (2008) says that the human resources (HR) departments of several companies maintain that most engineering graduates from India are 'unemployable'—though they come with good technical skill sets, they are unsuitable for working with international clients because of their 'culturally ingrained habits of subservience and passivity' (Upadhya 2008, 94). Another complaint from the industry is their lack of soft skills, which is a 'major weakness' (Upadhya 2008, 108). Within the bouquet of soft skills, engineering graduates from Tamil Nadu fare particularly badly in indicators such as 'learning agility', 'interpersonal skills', 'adaptability', and 'self determination' (India Skills Report 2018, 24).

The result is that many college graduates end up enrolling for advanced and/or additional degrees, taking up lesser jobs, or remaining unemployed while they wait for something worthwhile. With student debt to repay and lack of prospects, they end up being a huge financial and emotional burden on the family. They also face several disadvantages on the personal front as they are unable to get married as even rural families are known to be quite discerning in their search for a 'suitable match' for their daughters, a term that has come to preclude educational qualifications and an urban-based job. This desperation for work, personal crisis, frustration, waiting, and engagement in informal industry is often experienced as 'empty time' and 'temporal rupture' (Jeffrey 2010), often a source of severe mental anguish.

Reflecting this crisis, the promise of '100 per cent Placement' that most colleges advertise, hides as much as it reveals: only fifty per cent of a graduating class actually seek employment. The other half opt for further studies, join family businesses, or plan to get married and start a family. While a handful are able to hitch lucrative jobs that pay above Rs 14–15 lakhs per annum (about $17,000 US), meeting their expectations of a lucrative career, the average salary for a graduating batch is just Rs 4–5 lakhs a year (CCT Placement Records 2019; data available on website), less than the annual management quota fees for engineering college. Moreover, in colleges like CCT, prospective employers brought by the institutions to fulfil the promise of '100 per cent employment guaranteed' include banks, coaching centres, and business process outsourcing (BPO) companies, looking to fill roles that do not necessarily need an engineering degree. Students are also expected to take the first job they are offered, even if they are disinclined or overqualified for it. If they decide to let go of the offer, colleges sign off responsibility. When students question the fairness of such practices, the college echoes what the companies say: 'You are unemployable'.

What pushes candidates into the category 'unemployable' despite having an engineering degree? First of all, the term 'unemployable' emerges in reference to the IT services industry, in which workers need to supplement their technical skills with 'emotional labour' (Hochschild 2003 [1983]). Accordingly, a successful candidate must have an affable personality, good communication and soft skills, a well-groomed appearance and demeanour. In addition to doing business, they should be able to socialize with the client, make small talk, and be able to adjust to transnational cultural practices with ease (Brown 2003; Upadhya and Vasavi 2006, 25; McGuire 2013). These intangible factors have emerged as being crucial for securing employment in a highly competitive job market, where supply far outstrips demand.

The correlation between these desirable traits and urban location, upper class and caste position is so high, that Amanda Gilbertson (2018) advocates for recognizing 'cosmopolitanism' or 'cosmopolitan cultural capital' as an explicitly class- and caste-marked category, built on the genealogy of terms such as 'general category' that are similarly glossed (Deshpande 2013; Subramaniam 2015a). Despite the claims to meritocratic selection, studies have shown that terms such as 'merit', 'family background', and 'exposure' are the means through which biases of class and caste are worked into a secular language for screening candidates (Upadhya 2007; Subramanian 2015a; 2015b). Cases like Sugam's of one person beating the odds also affirm this: Sugam was a Nepali student, in CCT on a government scholarship given to higher secondary school top students in Nepal. Hailing from an upper-caste family, with educated parents running a successful business, Sugam attended some of the best schools in Kathmandu Valley before bagging the scholarship to come to

India. The scholarship also provided accommodation on campus and a monthly stipend of Rs 3500 ($42 US). His parents had always encouraged him to think of further studies abroad and India was the first in a range of destinations he had in mind for further education. He dreamed of a master's degree in the United States, after accumulating a couple of years of work experience. As Catherine had described him, he had a 'well-rounded personality', actively taking part in cultural events on campus, strumming his guitar as he sang popular 1990s songs, even striding down the catwalk for a modelling event.

Thus, more often than not, a candidate's class membership, regional background, and parents' education and employment status decide 'employability' and whether he/she can 'fit' within the globalized milieu of the urban workplace (Jodhka and Newman 2007; Fuller and Narasimhan 2006; Krishna and Brihmadesam 2006; Upadhya and Vasavi 2006, 46). Given that a third of the students at CCT are the first-ever graduates in their families and about two thirds come from backward class groups, these criteria for selection often work against their favour. This was true not only of CCT, but hundreds of colleges that were categorized as 'Category C' (non-elite).

To make up for these deficiencies, colleges aim to train students not only for technical competence, but also in good communication, personality development, and soft skills. Reflecting this, everyday life as well as the yearly cycle of the BTech programme is structured not only to inculcate expertise in students' chosen disciplines, but also to acquire the skills needed to become 'employable'. Students were expected to attend classes of forty to forty-two hours a week, spread over six days, comprising subject lectures, laboratory work, as well as classes in Professional Advancement and Career Enhancement (PACE), Business Etiquette, and Ethics. Special sessions on specific topics related to recruitment are also organized.[2] Such training, which borrows heavily from North American models, self-help literature, and New Age movement practices, is aimed at

---

[2] I had accompanied the students on one such session. Venkatraman, it was rumoured, had been paid ₹40,000 a day to train the third-year students in solving problems meant to test quantitative aptitude. A large man, he had an impressive moustache and had even more impressive ways of reeling off maths problems for students to solve without a single glance at his notes, and then recapitulating the same problems with the solutions in the exact same order in which he had first relayed them, but backwards.

the production of enterprising global professionals (Upadhya 2013, 2008; McGuire 2013). Such practices are also a result of companies asking colleges to bear the burden of 'employability training' (Upadhya and Vasavi 2006): colleges and even schools are under duress to 'inculcate cosmopolitanism' (Gilbertson 2018) through various programmes and activities.

Training to Get Ahead

The reputation of an institution is built on its ability to secure campus placements for its students. It is presumed by students and their families that institutions provide training to ensure they get placed, and that the 'big men' in the management have the necessary economic and social capital to ensure successful recruitment drives. Sometimes, sums of money were also exchanged and MoUs (Memorandum of Understanding) signed between colleges and companies, formalizing 'deals' which enabled companies to have the first pick among the outgoing batch of graduates. Generally, of the thirty-odd companies that visit for campus recruitment every year, one or two large IT service companies take the lion's share of recruits, while a majority of the others recruit just one or two students. For instance, in 2013 and 2014, a single company took eighty-four per cent and eighty-five per cent of recruits respectively. Though the company took recruits from all streams, the majority of candidates were from Computer Science, Information Technology, and Electronics and Computer Science. Because CCT was located in Tamil Nadu's textile belt, and its management comprised several 'big men' from the textile industry, the college also attracted several textile companies which recruited not only from the Fashion Technology class, but also from other streams, though their recruits remained few.

Despite the reliance on multiple rounds of tests and interviews to establish a seemingly meritocratic, above board, and transparent process, recruitment also comprises informal aspects: for instance, the probability of a company's visit to campus greatly increases if the equation shared by the management and the company's HR department is good. Moreover, every college has a dedicated Placement Cell, with a Placement Director (in the case of CCT, an erstwhile HR officer of an IT services company), who would use his contacts and 'influence' to attract prospective employers. Even if a company consistently recruited from one college every year, the relationship could fall through if the contact moved jobs. A discord or a split within the college management could also have a negative effect on recruitment drives. Though these aspects are not formally discussed, they were 'open secrets' among faculty and students with the tacit understanding that such relationships were the basis on which campus placements operated everywhere.

During recruitment drives, companies screen candidates for consistent academic performance (including school records) and through English language tests, group discussions, interviews, as well as through 'technical rounds' meant to gauge a candidate's numeracy, logical, analytical, and reasoning abilities. These tests construct the notion of meritocratic selection, even as students were eliminated based on factors such as their first language in school. Inevitably, the new recruits had a few traits in common: English medium schooling, urban exposure, educated and employed parents—traits which intersected with upper-class and caste categories.

However, by underlining these aspects, I do not mean to suggest that these factors become overarching determinants, with experiences and learning in college having no impact on a candidate's chances. As underlined earlier, various infrastructure is provided by the college to increase a student's chances of recruitment, and college is a transformative experience in ways that may not translate into the most lucrative job offer, but in the successful acquisition of 'exposure' and other positive attributes (Fuller and Narasimhan 2006; De Neve 2011; Zacharias 2013). However, following students in and out of college every day, the disjuncture between the formal 'training' with the students' lived experience seemed critical in the reproduction of disadvantage. Even though teachers and students are supposed to be committed towards improving graduates' employability by reproducing models of training in IT companies, they are also actively involved in the inculcation of locally sanctioned cultural and social norms. The engineering college emerges as a paradoxical site, producing both resistance and conformity to the prescriptions of employability training. While certain dispositions are learned, others are resisted and rejected. In effect, college life seemed to reproduce traditional norms of sociality and morality, while inculcating a positive outlook towards aspects of modernity such as technology and consumption, visible in the desire for mobile phones, western-style uniforms, and certain forms of cultural capital valued in a globalized milieu (Fernandes 2006; Deshpande 2003; Dickey 2013).

Learning a Professional *Habitus*

In Tamil Nadu, many engineering colleges are housed in large glass and chrome buildings that mimic the modernist campuses of large IT companies and Multi-National Companies (MNCs), trying to project a global look, while appearing disembedded from the neighbourhoods and towns in which they are located. The colleges also borrow from an urban–consumerist idiom for the nomenclature of their spaces—canteens are called food courts and cafés, kiosks selling beverages of different international brands are set amidst well-manicured lawns. Campuses are also equipped with facilities such as sports complexes, swimming pools, and air-conditioned hostel rooms. The campuses present a startling contrast to the surrounding pastoral landscape as well as stand in contradistinction to public universities which do not have similar luxuries (elite campuses such as IITs are an exception, in this respect). Viewed instrumentally, these numerous facilities are meant to justify the high fees charged, but they are also intended to cultivate the disposition required to traverse neoliberal cityscapes—a form of embodied knowledge that Bourdieu (1986 [1983]) calls *habitus*.

While this speaks to middle-class aspirations and the consumerist culture inaugurated by neoliberalization as well as the IT boom, what is peculiar is that modern amenities are combined with elements of Dravidian architecture such as spectacular ornate domes and columns, readily identified as a circulating signifier of Tamilness and 'neoclassical Dravidianism' (Bate 2009).[3] Perhaps to make this marriage of styles clear,

---

[3] I take my understanding of Dravidian neoclassicism from Bate (2009) as emerging from a Dravidianist paradigm which gained ground in the early twentieth century as a mode of political action dealing with vastly changing circumstances, including changes wrought by nationalism. In Tamil Nadu, this was characterised by the horizontal mobilization of lower-caste/class groups challenging the hierarchical paradigms of the highest caste/class, Brahmins. This involved a process of 'inventing traditions' (Hobsbawm 1983), generally seen as a part of the process of globalization. However, modernity is not just something that looks to the future, it has also come to be embodied in new things that appear old; things that look forwards and backwards and thus appears Janus-faced (Singer 1972, 400).

one of the colleges states 'Temple of Learning' in neon lettering on the building. The main administrative block of CCT, too, looks like a giant Chettinad house with clay tiles and iconic pillars, reflecting the caste affiliation of its management. Pictures of this 'iconic' building are printed on student notebooks as well as other publications with the tagline, 'An Endearing Shrine for Excellence in Education'.

The Dravidian neoclassic-inflected architecture and labelling of these buildings in language that denotes 'sacredness' seamlessly integrates with the heritage consciousness that has come to characterize cityscapes in Tamil Nadu today, combining 'neoliberal futurity and nostalgic longing' (Hancock 2008, 9). The state too, is invested in the idea of technological modernity that blends seamlessly with the political parties' espousal of neoclassical Dravidianism—both aural and visual (Bate 2009; Pal 2019). As Sanjay Srivastava (2006, 32) observes of modern constructions that attempt to imbibe traditional features, this is a modernity that involves juxtaposition and recirculation, where the economic and the cultural reinforce each other. Drawing from the organization of interior and exterior spaces, it seems that colleges are imagined as spaces producing modern Tamil subjectivities: spaces where students learn to be *decent* middle-class subjects (as opposed to *local* with its connotations of being lower class and caste), who are produced to be culturally loyal even as their education ostensibly trains them to be 'global' employees.

This is also manifested in the dress code that mandates not only what kind of garments are to be worn on campus, but also exact style and cut. Journalistic critiques of dress codes in these colleges have discussed their regressive implications and infringement upon personal choice. Others have considered the formal endorsement that such rules have lent to 'shaming' discourses, especially those that pin harassment and assault on women's sartorial choices (Phadke 2007).

However, the dress code is not limited to managing students' sexual awakening, but has layered meanings. The caveats mentioned in the dress code also seek to achieve professional goals by steering youth away from local styles. Thus, in addition to forbidding sleeveless clothes and short tops, young women are mandated away from hyper-feminine styles such as shimmering *anarkali* kurtas and *langa voni* (half saris) with links to traditional local culture such as puberty rituals. Instead, the dress code urges them to adopt a style that is simpler, more modern, Westernized, and gender neutral (such as kurtas and jeans, and the trouser suit). Similarly, male students are forced to distance themselves from youthful clothing such as brightly coloured clothes that communicate 'style' (Nakassis 2016) to cultivate more adult-appropriate clothing. [4] Students are also barred from wearing articles of clothing such as black shirts, which is linked to the Dravidian movement. In other words, they are encouraged to cultivate an appropriate apolitical distance from local politics to settle into a 'global' career. Given the college's provincial location, it is perceived that students must be coached into making these appropriate fashion choices that would, in turn, help them adapt to a globalized professional setting, and develop suitable 'taste'.

Also reflecting this, the prescribed CCT uniform for students is a navy-blue trouser suit with the college insignia embroidered on the left-hand side, the same for men and women students. Therefore, even though women students are not allowed to wear trousers and short tops/shirts on a regular basis, an exception is made for the uniform in order to appear as desexualized and competent professionals. The students also appear for campus recruitments in the uniform, reflecting the ways in which the trouser suit has become the de facto formal attire to attend job interviews, even though students wear Indian formal attire such as saris for other formal events such as convocation ceremonies. Apart from the job interview, the uniform is reserved only for special occasions like college annual day, association meetings, and class photographs, when students need to present a formal self. Even at these events, female students do not always wear the full trouser suit, sometimes just slipping on the blazer over their salwar kameez and dupatta, reflecting the 'balance' between being respectable and becoming professional.[5]

---

[4] See Nakassis (2010; 2016) for a detailed discussion of counterfeit brands and their style quotient among male youth in Tamil Nadu.

[5] Women teachers also mimicked the same style: many of them wore a lab coat (with college insignia; the teachers' uniform) when they come to teach in class, even during the summer months, in order to present a professional image. Teachers said that this was a standard practice all over Tamil Nadu; a teacher had worn such a lab coat in her previous college too. Students in the mechanical stream had to wear a lab uniform, a pair of dark blue overalls, when working with welding machines and lathes, in the lab with a pair of sturdy black shoes. The women, numbering only a handful in the mechanical stream, changed into these clothes just before entering the lab, and avoided wearing the shoes, for which they were reprimanded. The uniform prescriptions were therefore enforced very seriously in the college.

The prescriptions above are not exceptional, but a common feature of colleges in South India (*Daily News & Analysis* 2015; *The Times of India* 2013; Lukose 2009). In cultivating students' identity through a professional style, colleges are able to reiterate segmentation from other globalized industry such as textile and knitwear export and the 'menial' aspects of the engineering profession. Through its architecture and the prescription of the uniform, the engineering college reiterates its identity as a site of display not just of new-age technology and spaces, but also of new-age careers involving membership of a professional class. The symbols of work in the globalized IT industry start from the college itself.

However, a close examination of students' everyday life shows the subterranean anxieties that undermine such aspirations. The rules, and students' responses to them, reaffirm the principle of conformity, and the discomfort of adopting flexibility in transnational settings. While aspects such as the uniform are reified as a necessary part of becoming a professional, any deviance from the norm invites scrutiny: for instance, foreign students enrolled at the college often complain that their casual attire for outings draws scrutiny from hostel wardens and security guards. 'Just jeans and T-shirt mean so many things here', one student said, after the security guard reported to the principal that he had seen her step out in Western clothes and the warden called her to check her whereabouts. The adherence to local norms as well as the continued hyper-sexualization of bodies that were clothed differently, strengthens an existing rigid moral stance against cultural difference, without the realization that these attitudes also contributed towards unemployability.

Moreover, the emphasis on developing a flexible 'global' identity contradicted local discourses of Tamil identity and language which are ubiquitous in Tamil Nadu as a diffused effect of the Dravidian movement (also see Nakassis 2016). The movement, after the 1940s, diluted its rationalist stance and its opposition to brahminical hegemony and caste hierarchy. Instead, the movement promoted an anti-Hindi, anti-North Indian stance and emphasized a common Tamil identity based on what has become the essentialized 'Tamil' values of female chastity, male valour, motherhood, and love of the Tamil language (Barnett 1976). By defining gendered values as Tamil and spreading them through cinema, the Dravidian movement not only developed a broad base of political support that overshadowed communal and caste differences, but also engendered a heightened sense of Tamil identity (Pandian 1992; Lakshmi 1990; Vera-Sanso 2006).

In CCT, this sense of identity inspires the reverence and speaking of Tamil whenever possible, and the construction of Tamil as the basis for a 'moral community' of certain values not shared by 'outsiders'.[6] The boundaries of this community were marked by terms such as 'Peter' or 'NRI' (Non-Resident Indian) which were often used informally as a rebuke or derogatory reference to someone who spoke consistently in English in tandem with what were perceived as other signs of 'elite' behaviour such as wearing Western clothes, drinking, smoking, or visiting a pub. Such attitudes become particularly accentuated vis-à-vis the 'NRI' students at the college, who represent a globalized population. This term did not necessarily mean 'Non-Resident Indian': it was an elastic term that had regional, cultural, as well as class connotations. I was often referred to as 'NRI', even though I come from the neighbouring state of Karnataka, had gone to undergraduate college in Chennai, and have working proficiency in Tamil. However, my clothes (cotton salwar, kurta, and stole), short hair, tendency to slip into English while speaking, departmental affiliation in Delhi, and the privilege of being able to spend a year hanging out in their college betrayed a different class and cultural position.

The term 'NRI' also encompassed international students from South Asian countries such as Bhutan and Nepal and African countries such as Sudan and Somalia. While the South Asians stayed at the 'NRI Hostel', the African students were not even provided accommodation within the college for 'moral reasons' as a non-teaching staff member phrased it. Admission was also closed to female students from African countries in CCT for the same reasons, although many other colleges in Salem admitted them. Functioning under a separate dean, male and female students in the NRI hostel live in the same building but in separate wings, have a separate food menu, and can venture out later than others. 'We have to provide them these concessions because they are from a different, modern culture. Not Tamil', the dean for 'NRI' students said.

Words such as 'NRI' and 'Peter' are not merely slang circulating among students (cf. Nakassis 2016, 105), but are also used in formal speeches.

---

[6] This mistrust of the 'outsider' is captured in the use of the word *'paradesi'* as a derogatory term among the students.

Several guests, while presiding over events at CCT, advise the students to work hard, study well, and get good jobs, 'but never turn into Peters'. One guest, a short film director, likened learning Tamil to nurture by mother's milk and said that just as breast milk ensured immunity and good health throughout a person's lifetime, the mother tongue ensured immunity for one's moral life. The scalloped ways in which 'Peter' emerges as a marker of linguistic, moral, and gendered 'excess' points to the discomfort with foreign ways and manners. These discourses also show the fashioning of a self that is rooted in the local, in an education system whose premise, ironically, lay in mobilizing a globalized labour force. However, despite the negative moral, social, and cultural value attributed to English in everyday life, the learning of English was emphasized to increase employability. Such 'practical romanticism' (Herzfeld 1997) invigorates academic life in the college, while social life continues to be marked by disjuncture.

'Culture Shock'

The importance of English in increasing employability is highlighted almost every day in the college. It was implicit that English even overtakes subject knowledge during recruitment as it is highly valued in the workplace. This was even stated by two of the college teachers in an article published by them, in an indexed journal:[7] 'students who are the future employees have to deal more with soft skills than with actual knowledge about particular situation because customers appreciate an employee who is willing to help and listen to the complaint ... Hence, training the students in soft skills has become the main agenda in colleges'. Accordingly, the college offered language lessons marketed as 'Business English' or 'Communicative English', in which exercises are designed to help students learn corporate etiquette ('develop a strong handshake', 'if you are meeting a woman for business, let her extend her hand first', 'don't stand too close to a lady in the elevator'), in addition to general communication skills such as how to compose an email, conduct a business meeting over the phone, and so on. It was evident from the way exercises were designed

---

[7] Citation details not given to protect the identity of the authors, who were my interlocutors.

that the objective was to develop an employee who could fit comfortably into companies which have 'offshore' business models.

The turn to 'Business English' from traditional (colonial) models of teaching English through literature has been widely embraced in colleges across Tamil Nadu, and perhaps even across the country (Tharu 1998), as a necessary course correction to individual and collective 'lacks' in the job market. These attempts at socialization into the job market have been validated by various international testing agencies such as Cambridge and the British Council, which not only supply material in the form of textbooks and audio-visual software for workplace simulations, but also train teachers and provide additional certifications to students. The model has become so popular that there is now a veritable industry—both formal and informal—around the teaching of English. The AICTE too, recommends that this model be adopted by engineering colleges, and its Model Curriculum (AICTE 2018) prescribes a syllabus that focuses on everyday conversations in the workplace, giving interviews, making formal presentations, and improving pronunciation by removing mother tongue influence. However, as the following excerpt from my field notes shows, such exercises played out in a complex fashion in the classrooms.

30 April 2014
Salem
I sat through an English class today by Dr Z, the head of the Department of English at the college. The activity for the day was Role Play, a part of the Speaking Unit. Dr Z described the situation to the class: 'You have subscribed to a new plan on your mobile phone, but your Service Provider has not yet activated the plan. Call Customer Care to find out the problem, and sort it out'. She then asked for volunteers to role play the situation in front of the entire class. Because no one put their hands up, she picked two students at random. Sejal would play the role of the customer, and Krishna, the customer care representative. Krishna immediately stood up from his place and walked to the front of the class. Sejal, on the other hand, stood up bashfully and stood rooted to the spot. In a shaky voice, she said that she was too nervous to participate in the role play. When Dr Z insisted that she come, she said she was recovering from a sore throat and would not be audible to the rest of the class. At this, Dr Z began to chastise her in front of her classmates

for being too shy. A visibly shaken Sejal then started making her way to the front of the class.

Krishna: Welcome to Vodafone, madam, how may I help you?
　class giggles
Sejal: Hello ... (soft and stammering), where is the plan ...? why haven't you activated my plan?
　wiping sweat off her upper lip, and covering her mouth with a handkerchief
Krishna: We will activate the plan immediately, Madam. Sorry for the inconvenience.
Sejal: Okay. Thank you.
Krishna: Have a good day! Bye!
Sejal: Okay.
　Running back to her place.
　Class giggles again.

Dr Z provided some feedback: 'Sejal, why are you so shy? Please be clear next time, explain the problem'. She then instructed the class to role play the situation with the person sitting beside them. 'Practise once or twice', she says, before sitting down on her chair.

Discussing the activity after the class in private, Dr Z said that in fact Sejal had very good communication skills and spoke English very well. 'She was uncomfortable only because she had been paired with a boy. That is why she was resisting so much. If I had paired her with a girl, the response would have been totally different', she said. 'Girls are much more articulate and expressive within their own groups', she continued.

As is evident in the note above, such exercises are highly uncomfortable for many students, who are too self-conscious to assume different roles and personas in front of a class of fifty or sixty students. To stand in front of a class and speak in English with a student of the opposite sex goes against the norms of everyday sociality which emphasize non-hypervisibility, especially for young women. While Krishna was able to sail through the exercise through the adoption of an exaggerated comic persona, such options were not viable for women. If Sejal had engaged in a free-flowing performance of conversation as the exercise warranted, she would have been teased later for being a Peter, for being 'too free' with Krishna, and for putting on a scene.

This disjuncture between everyday norms of sociality and the prescriptions of employability training was part of what was termed as 'culture shock' on campus. This was not limited to such exercises but seeped down to different levels of interaction in everyday life. For instance, everyday life in CCT emphasizes hierarchy between student and teacher, parent and child, and seniors from juniors. In fact, teachers in the college drew from a familial idiom of speech, addressing the students as 'child' or even '*thambi*' (younger brother). In response, students addressed the teacher with their heads down and shoulders hunched in deference with 'Sir' or 'Ma'am'. Across different colleges in Tamil Nadu, there have also been controversies related to individual administrators and teachers who enforce discipline with an iron fist. Hundreds of students have claimed to be victims of verbal abuse and physical assault by college authorities for acts of indiscipline ranging from leaving shirt buttons open, talking to members of the opposite sex, or not doing well in academics. These regimes of fear neither help students develop the confidence to assume a professional stance in the workplace, nor help students to cultivate a direct conversation style with people in authority (such as using first names for one's superior), as required by the global services industry.

Moreover, despite a concerted effort by the college to make students comfortable speaking to members of the opposite sex in simulated 'professional settings' such as workshops and during class exercises under the vigilant eye of teachers—such interactions remained awkward and ridden by anxiety, as the classroom exercise recorded above shows. Therefore, even though the aim of certain exercises was to train students to have an embodied confidence to engage in such interactions, the milieu of the college required them to behave differently and seemed detrimental to developing an amicable rapport between members of the opposite sex.

Pedagogical emphasis was also markedly different from school, where engineering college aspirants usually focused on the science subjects (physics, chemistry, and mathematics), almost to the neglect of languages (also see Sancho 2013), with intense coaching in these subjects, starting as early as class nine. In what seemed almost a paradoxical turn of events, the training in engineering college emphasized English classes, and even the need for foreign languages such as Mandarin, German, and Japanese in order to secure jobs. This feeling is augmented by a sharp gap between students' existing proficiency in English and the college's use of American and British resources to teach English. Study material include comprehension exercises of reports in British and American newspapers and the use of recorded speech in different accents in the language labs, so that students can learn to comprehend accents of future customers. The educational software used in the language labs are intended for Western learners, but have been superficially adapted for Indian students. For instance, the illustration videos feature white people in Western contexts but their names amended to South Asian-sounding names. The videos also make extensive use of terms and experiences from urban elite settings such as malls and restaurants, and presume a certain familiarity with such locations. For instance, one of the introductory videos said that placement visits by companies are like shopping trips, and the interview process is like a trial when companies decide if the new resource would 'fit'. The accompanying visual shows a changing room in a departmental store, and a woman entering it carrying a set of clothes.

Such presumptions are ubiquitous in exercises too. For instance, a role-playing prompt requires students to pretend they have just returned from a meeting in London and asks the students to 'discuss the possibility of starting a branch in Hereford' without any cultural, geographical, or social context about Hereford provided. Thus, the material focused on commercial and business contexts to which the students had no prior exposure. When asked about the possibility that these illustrations were out of the average student's frame of reference, one of the teachers said that they are preparing students for international exams such as GRE, TOEFL, and IELTS.[8] 'Students here do not read anything else. We want them to get "exposure". The students from rural backgrounds do find it difficult, but we cannot lower the standard for 20 pc of the students'.

The elite orientations and backgrounds of teachers exacerbates feelings of alienation in the classroom, and there can be a large gap between the life-worlds of disadvantaged students, their teachers, and the globalized worlds presented in their lessons. Moreover, there is underrepresentation

---

[8] Members of the faculty of the English Department were trainers for programmes such as BEC, GRE, and TOEFL.

of disadvantaged groups such as Scheduled Castes and Scheduled Tribes among faculty members in many institutions, including elite institutions such as IITs. Private universities and colleges consider themselves exempt from having to follow reservation policies in recruitment of teachers—forty-six per cent of unaided private universities and seventy-six per cent of unaided private colleges do not follow reservation or any other affirmative action policy in their recruitment (Venkatanarayanan 2019).[9] Teachers from elite backgrounds who reproduce modules of employability training and its dominant ideology may not be able to help students help themselves, and may even prescribe learning material out of the grasp of disadvantaged students leading to what students called 'culture shock'. For instance, one of the self-help books prescribed for the first-year class is *The Power of Positive Thinking* by Norman Vincent Peale, a 1952 American bestseller. The book begins by introducing its premise that many of life's failures are results of a vicious cycle of negative thought resulting in an inferiority complex, which could be remedied through 'the power of positive thinking'. This particular abridged edition for young people was brought out by the author as 'nothing could happen to a person of greater fortune than to master the positive thinking technique early in life' (Peale 1992, iii).

'What is an inferiority complex?', one of the students asked within the first five minutes of the books being circulated. (The first line of the book is, 'Inferiority complex creates barriers in our personalities' and 'is a malady that arises out of the misty past in the dim recesses of our personalities' [Peale 1992, 3]). 'It is the feeling some people get when they feel they cannot perform as well as others', Mrs V., who was supervising the class, replied. She continued walking around the class, clearing other doubts students had. At the end of the class, when the teacher asked for feedback on the book, two young men responded saying that they did not know they had an inferiority complex. The teacher responded that it is good to identify this problem, because now they could start using 'the power of positive thinking' to get over it.

When the teacher asked how many pages they had been able to complete in an hour, most students said they had been able to complete only about five to six pages. The difficulty of comprehending the terms and

---

[9] See this report in *The Wire*, https://thewire.in/uncategorised/tamil-nadu-reservation-quota.

reading much of the book demonstrates the incompatibility of the resource material with readers' abilities. My own perusal of the book showed that the book addressed not just a Western subject, but was written for a Protestant[10] readership from America of the 1950s, replete with phrases from a milieu that would be unfamiliar to students at CCT. A cursory search on the Internet even showed the book itself was highly controversial and had invited criticism from scientists, mental health practitioners, and theologians as it prescribed an unethical method akin to self-hypnosis. The teachers were, however, unaware of any of this, and believed that self-help books such as *The Power of Positive Thinking* fulfilled two objectives: one, enhancing vocabulary and awareness of the outside world; and two, teaching students that 'self help is the best help', an oft-reiterated mantra, especially in reference to changing oneself to suit the workplace atmosphere.

Students' perception was often different. For economically disadvantaged students who came from Tamil-medium or government schools, 'culture shock' was the advantage enjoyed by their elite classmates, who through their fluency in English as well as engagement with forms of popular culture such as English films and sit-coms, were better positioned to engage with learning material. Such feelings of not being able to converse easily in English or understand American or British pop-culture references were made worse through labels such as 'lazy' or 'rural background' becoming attached to them. In such a context, students' silences, pauses, and reluctance to respond vocally highlights the continued 'self-marginalisation of the subaltern student' (Jaware 1998), as a response to their experiences and backgrounds being sidelined in the classroom.

However, marginalized students rework the meanings of their own positions through their expressive practices by clarifying what it means to have 'real talent' or 'real merit' in the classroom. Hence, students might remake the meanings of what it means to be a 'knowledgeable self' (Luttrell 1996) even as they might internalize labels of failure. An impromptu performance put up by students of the special 'Bridge Course' for Tamil-schooled students highlights the different dimensions through which students perceive their inequalities. In the skit, one of the characters, Aravind asks for help from his friend Kalaiarasu to learn maths.

---

[10] Peale was an ordained minister of the Methodist Church.

Kalaiarasu readily helps him, and Aravind is able to do well in his exams. Sometime later, there is an announcement by the English Club about a round of competitions for literary competitions such as 'Just a Minute',[11] essay, and debate. Keen to participate, Kalaiarasu asks his English-medium educated classmate Aravind whether he can help him, but Aravind turns him down saying these competitions are very tough, and there is no chance that a Tamil-medium student will be able to participate in these competitions. He goes on to be the only participant in the competitions and collects all the prizes. The scene then cuts to the alternative and more desirable scenario: one in which the English-medium student helps the Tamil-medium student to prepare for the competition and they both share the prize offered in the competition.

Given the disadvantages faced by Tamil medium schooled students, it is interesting to note how the linguistic backgrounds of the two characters plays out in the realm of the ethical. Kalaiarasu's Tamil-medium school origins (with class and rural background connotations) is given a positive spin with remarkable knowledge of mathematics and helpful nature, indexing his culture and merit, while the English-speaking character emerges in opposition as a self-serving, elite individual who wins only because of his language capabilities, and lack of competition from his Tamil-schooled peers.

The skit must be read in the light of the 'triple challenge' that students from marginalized backgrounds face while entering higher education (Vasavi 2006): one, the inheritance of diminished self-worth as a result of persistent humiliation; two, deprivation and indignities because of the non-recognition of the traditional skills, economy, and other knowledges that they tend to possess; three, the celebration of urbanized upper-caste youth as 'meritorious' or 'talented' because of their engagement with new Western global cultures. However, when characterizing themselves, they show a positive picture that highlights achievements in technical subjects such as maths that does not need English language proficiency. From their point of view, knowledge of English seems hollow, valued only for the performative confidence it enables its speakers to have, rather than

---

[11] Based on the popular BBC radio comedy panel game, where contestants have to talk for a minute without hesitation, repetition, or deviation. The game calls for a good command over language, knowledge about diverse subjects, and confidence.

actual excellence. The skit also shows resistance to the high economic value placed on the language, accent, and behaviour of the urbanized upper caste/class in workspaces that deal with business processing and IT (McMillin 2006, 240; Singh and Pandey 2005; Krishnamurthy 2018), and its reiteration in the college. The scenario in which the English-medium student walks away with the prize should be read in the context of the employers' bias for English-speaking students. The teachers, too, had constantly repeated in class that top IT companies recruited only students with good English communication skills, while students who were only fluent in Tamil/regional languages would be left in the lurch, looking for jobs in the smaller IT companies or joining the informal sector.

Such sentiments are also expressed in movies such as *Kattradhu Thamizh* (Ram, 2007), in which the protagonist Prabhakar (actor Jiiva) is a dedicated Tamil teacher from a rural area, now living in Chennai. His profession is shown to be not just a job, but a calling. However, in the context of the early 2000s, when engineering colleges and software parks were mushrooming in Chennai's outskirts, benefiting from sops doled out by the government, Prabhakar feels 'left behind' with his MA in Tamil and school teacher job. He is constantly belittled and teased for not possessing the 'right kind' of cultural and symbolic capital. His students do not listen to him; several people exploit his gentle ways. These acts disgust and disappoint Prabhakar. The meagre value placed on his education also ensures that he cannot fulfil any of his life's dreams (such as marrying the woman he loves, who is forced into prostitution because of poverty). Compounded by these everyday humiliations, he sees a clear dichotomy between 'those inside Spencer Plaza,[12] and those outside it, those inside Satyam Cinemas and those outside it; those inside an ATM and those outside it', and blames the IT boom for his misery. The rest of the movie is focused on a vigilante trope, in which Prabhakar transforms from a gentle soulful teacher to a cold-hearted murderer who is keen on ridding the city of elements against 'Tamil culture'.

The scenarios in films such as *Kattradhu Tamizh* (Ram, 2007) and the more recent *Vellaiilla Pattadhari* (Velraj, 2014) resonate closely with the situation in Tamil Nadu, and in India broadly, where unemployment is at

---

[12] A popular mall in the early 2000s, situated in the heart of the city of Chennai. Satyam Cinemas was one of the first multiplexes in the city.

an unprecedented high. The films dramatize the unemployment crises by showing how many of the unemployed have been pushed into the category of the 'unemployable' despite holding advanced degrees, as they lack good communication skills in English and other soft skills valued by the global industry. Without examining the objective reasons for this global slowdown and retrenchment in capitalist processes (see Upadhya and Vasavi 2006; Saith and Vijayabaskar 2005; Nisbett 2009, 26–47; Upadhya 2016), the films focus on subjective factors such as the elitism of the new IT class alone leading to what may be termed as a backlash against IT culture.[13]

Such a discourse against 'IT culture' is widespread among the faculty and college employees in their everyday conversations. For instance, the head of the Placement Cell, Mr BM spoke bitterly of the 'moral problems' in the corporate sector and characterized a corporate job as 'being after money' with no concern for what is morally right. Because of that, he had even quit his job and returned to his hometown Salem, he said. Ironically, his job in CCT involved persuading his contacts in the corporate sector to come and hire at the college.

Students echoed similar sentiments about the IT sector: many interlocutors in the Civil, Textile, and Mechanical Engineering streams said that they and their families had based their non-IT education decisions on a critique of the globalized culture of software/IT companies as being 'too Westernized' and 'having moral issues'. Such subjectivities counter the idea that the 'IT craze' had completely seized the popular imagination of youth in South India (cf. Nisbett 2009). As a result of the tough competitive market as well as the changes in lifestyle required by the IT industry, not having an IT career has become a matter of pride and assertion in Tamil Nadu, where some cars and bikes are stamped with stickers that read 'Civil Engineer' and 'Mechanical Engineer' as badges of pride. These stickers index an older, albeit slower, path to middle-class status and respectability through traditional engineering disciplines, which are still dominated by older forms of (masculine) capital and public-sector funding (PSUs). The imagination of the civil engineer and mechanical engineer is not located within the glamorous office of the multinational company, but to bricks and mortar, machinery, hard labour, and an older

---

[13] For a more detailed discussion on *Vellaiilla Pattadhari*, see Hebbar (2018, 2020).

version of development. Implicit in the valorization of these traditional disciplines is the contrasting figure of the globalized software engineer, whose labour is disembedded from field, site, and nation, and whose sole motive seems to be profit. Accordingly, multiple critiques of 'IT culture' were repeated by students and teachers alike: they complained about time zones intervening in family life, the 'moral' standing of IT workers travelling on-site, and the challenges presented by IT work to traditional and gendered models of productive and reproductive labour. Anxieties stemmed from postings in different cities, the unpredictability of the next 'site' of work, and the threat presented by proximity to machinery and technology to the natural rhythms and rhymes of the body ('cannot sleep at the same time everyday', 'IT workers have problems getting pregnant', 'I will be too far from my hometown').

'IT' is also synonymous with feelings of alienation and anonymity as shown in several films such as *Kattradhu Thamizh* (2007) elaborated earlier, and more recently, *Maanagaram* (Kanagaraj, 2017). In *Maanagaram*, which means the 'metropolis', the protagonist from Trichy (whose name we never find out, an allusion to the anonymity of the thousands of nameless, faceless migrants to the metropolis), comes to Chennai for a job interview and finds the city a merciless one. He gets beaten up, not once but twice in cases of mistaken identity. In what may be seen as a literal manifestation of the unemployment crisis, his certificates and degrees are also cruelly snatched from him. The kernels of scepticism and doubt that are contained in narratives from the engineering college find embodiment in these films, which characterize the globalizing city as one filled with apathy and danger. Though the resolution of *Maanagaram* (2017) is far more optimistic than the grim *Kattradhu Thamizh*, it remains steadfast in its criticism of the globalizing city and increasing apathy of a disembedded, floating population.

Like the discordant notes against employment in IT jobs in other aspects of everyday life, some of the learning material also carries grim warnings about the possible pitfalls of seeking career fulfilment in corporate life. Shaped by historical and sociopolitical discourses as well as contemporary contexts, these warnings are most evident in textbooks authored by the teachers from the college. For instance, a textbook authored by teachers from the college entitled *Professional Ethics and Human Values* showcases a narrative that runs counter to the perception of colleges reproducing the same self-regulating discourses as software companies. Despite the stated course objective of developing 'moral thinking in professional practice', the course reiterates a rigid and traditional relationship with authority, likening employment to classroom and military contexts (*Professional Ethics and Human Values* 2014, 45),[14] even though it was well known that IT employers prefer an egalitarian style of working, and lines of hierarchy and authority often blur (Upadhya and Vasavi 2006). As in the illustrations earlier, local contexts and ethics are given more priority than flexibility in transnational settings.

'Ancient empires have been replaced by MNCs' (*Professional Ethics and Human Values* 2014, 158),[15] the book proclaims, sounding a cautionary note on Multinational Corporations (MNCs), and the fact that developing countries offer multinational corporations a competitive advantage:

> Developing countries offer a wide market for the products of MNCs. Often MNCs dump products and processes in the name of technology transfer. Most of the MNCs create their policies and issues and often they use developing countries as a ground for experimentation. One can write a thesis on the pros and cons of MNCs in the developing countries ... Workers in the developing countries might be willing to take relatively greater risks. More often, this willingness is due to their poor living conditions and ignorance. MNCs often exploit them. MNCs must respect the basic rights of people in the country where they do business. They must eliminate any risks in their operations as and when they can, while still making reasonable profits. Workers must also be compensated suitably for the risks. (*Professional Ethics and Human Values* 2014)[16]

Such notes of warning seem to strike a counter narrative to the training programme, which is oriented to making students 'employable' in such companies. Perhaps this is also a tacit acknowledgement that despite their best efforts, the prospect of finding employment was a difficult one, perhaps not worth the many transformations it warranted.

[14] Full citation details not provided to protect the anonymity of the teachers and the college.
[15] Full citation details not provided to protect the anonymity of the teachers and the college.
[16] Full citation details not provided to protect the anonymity of the teachers and the college.

However, Tamil Nadu's advanced affirmative action policy that reserves sixty-nine per cent of places for various disadvantaged groups has been important in ensuring that several communities have been able to successfully overcome their historic disadvantages through higher education in streams such as engineering and have successful careers in IT, and even migrated abroad in large numbers. This includes not just 'creamy layer' castes within the OBC fold, but also deprived sections such as Most Backward Castes (MBCs) and Scheduled Castes (SCs). Nowhere is this more visible than in urban pockets of Kongunadu such as Coimbatore, which are considered retiree communities, thanks to youth leaving en masse to metropolitan cities like Bangalore and Chennai, and overseas for IT jobs. This is also an indicator that despite the many challenges, large populations can successfully transition to engineering jobs and become part of global communities. Therefore, while it is possible to ascertain such critiques of the culture of IT employment from these localized voices, one must also evaluate possible positive effects of these programmes on marginalized youth.

The Remaking of Self and Subjectivities

In an essay on her experience conducting skill development courses for students from disadvantaged groups, Usha Zacharias (2013, 319–325) argues that soft skills programmes have the potential to psychologically transform marginal subjectivities caused by class, caste, and regional positions, though they ignore historical antecedents, economic deprivations, and social structures: 'The psychologisation of marginality, while it leaves historical and structural reasons for marginality untouched, empowers the students to tackle the problems of selfhood formation in teenage year(s) in a particularly effective manner' (Zacharias 2013, 319–320). Zacharias argues that the effectiveness of soft skill programmes in enhancing the understanding of the self makes students aware of the differential values given to language, culture, and expressive traditions, and hone themselves accordingly. Such cultural expressions show student resistance to dominant ideologies at its richest: they demonstrate how such conflicts are regularly expressed, and more importantly, partially constructed by, everyday practice and discourse (Lukyx 1996, 255).

A few months after watching the skit about the conflict between the Tamil-medium student Kalaiarasu and English-medium student Aravind, I revisited the class that had enacted it. On the insistence of their English teacher, I asked the class if they would like to tell me whether the English class had helped them and how. At this, the class began to shout in unison, 'SAM, SAM, SAM ...' Since I did not understand what was going on, I looked at the teacher for help, and she pointed at a young man who occupied the last bench in the class. 'Sam' stood up, and the teacher introduced him: 'Palanisamy', she said, 'has requested his teacher and classmates to call him "Sam" inspired by the skit'. The young man stood grinning. The teacher continued that he was determined to speak better English and become 'Sam'. The whole class burst out laughing, and the teacher too joined in the laughter. Sam did not say anything, but looked shy.

Palanisamy's strategy of changing his name reflects hope and aspiration to 'fit in' the globalized framework within which he hopes to find employment and success. It represents a desire for a shift from a local name carrying regional markers to a name that appears universal and globalized. Such change of names is common in the processing outsourcing industries where South Asian employees often must adopt names that are easier for their Western clients to pronounce, in addition to becoming familiar with Western phrases and consumer practices (McMillin 2006; Singh and Pandey 2005; Krishnamurthy 2018). Such transformations are a common result of skill development programmes. As Usha Zacharias (2013, 325) mentions in her case study using the example of young women who usually 'outgrow' half saris and wear jeans by the end of the soft skills courses, such transformations are common. In other words, students assimilate the larger ideology of soft skills training that becoming professional or crafting an employable self requires a Goffmanian 'self presentation' that includes theatrical techniques such as speaking in different accents, taking on a distinct character, and wearing a 'clothing of skills' (Zacharias 2013, 325).

While Palanisamy's case presents a direct correlation between the curricular framework and the remaking of cultural subjectivity in accordance with the imagination of the global, such attempts at transformation show that struggles are not limited to curriculum but exist at the level of self and subjectivity. These are also aided by the rhetoric of 'self help' and identity transformation that permeates life in engineering college. From the books prescribed for reading in English classes to interpretations of popular fiction such as Harry Potter and films such as *My Fair Lady* (1964), students are forced to acknowledge in every narrative, the underlying fable that men and women will benefit from learning and transformation if they are able to make the self the locus of change, even under trying circumstances. Learning material, whether in English, Personality Development, or 'Engineering Ethics' classes, is honed to fit this dominant ideology, as prescribed by prospective employers.

However, the question remains whether just the remaking of marginal subjectivities and the injection of hope and confidence in young people through inculcation of the dominant ideology and self-help is enough to make a difference at interviews and increase a student's employability. Official sources at the college said that companies often insist on weeding out CVs of Tamil-medium schooled students before the interview stage. But as I have mentioned in the previous chapter, lobby groups of self-financed colleges have been pressurizing companies to change their recruitment policies to factor transformations brought about by processes such as English learning and soft skills training at the college.

Conclusion

The chapter shows the enormous pressure cast by concerns of employability in shaping everyday life at college. The proliferation of colleges as well as the requirements of the IT services industry has shaped the way engineering education is experienced primarily as the need to emerge 'employable' not just through the accumulation of technical knowledge but also by inculcating good communication and soft skills, and imbibing techniques of 'self-help'.

Curricular frameworks as well as teaching methodologies emerged as tedious disciplinary projects to train students into becoming 'professional subjects' who would be comfortable dealing with foreign clients and neoliberal horizons, strengthening existing forms of cultural capital to the detriment of those with disadvantages. Critiques of IT work culture arose from all quarters, even as they strove to assimilate the dominant ideology. The result, therefore, is a 'partial penetration' (Willis 1977) of these discourses that allow students to refashion their identities to show aspects of being 'professional' and upwardly mobile, while distancing themselves from lower class and rural identities. The resultant disjunctures are important in pointing to the ways that students also imbibe qualities that could be detrimental to their job prospects.

However, despite the resources made available for students to emerge competitive in a global scenario, the lived experience of students remains tied to local frames of reference such as respectability and the discourse of Tamil morality, which led to a discourse against what students see as the 'excesses' of speaking in English, dressing in Western wear (for women), and portraying too confident a demeanour. Despite what they taught, teachers and other resources that are locally authored also critique global workplaces, its culture, and morality.

Many of these discourses have their basis in the extremely gendered milieu in which the college is located, playing a key role in every aspect of life in the college from enrolment to the negotiation of everyday life. In almost every aspect of college life, we see students cultivate a sense of 'respectability' by demonstrating abhorrence for the 'sexual' and the 'risky', reiterating their adherence to local moral discourses (see chapter 5 in this volume for an exception). Though 'employability training' attempts to rise above these 'local' markers, college life is shaped very much by the biases and prejudices of the key actors, including the college managements and the ways in which they balance their perception of the needs of the global market with their own 'reputation' in the local prestige economy. Situating engineering colleges in this gendered milieu anchors the celebratory accounts of women's unprecedented entry into streams such as engineering and the software industry, drawing attention to both new forms of 'feminization of labour' as well as domestic arrangements necessitated by engineering degrees and their perceived social value. To focus on the gendered context of the college and women's negotiations within these spaces is to draw attention to the ways in which women's labour is valued, and the agentive reworking of tropes such as 'respectability', which have held long associations with women and labour, into every aspect of women's lives from their chosen academic disciplines to personal friendships.

4
Manufacturing Respectability
Gendering the Engineering College Boom

'If we didn't have so many problems, we might have been sending satellites to Mars.'

Monika, student, CCT

Dressed in nylon salwar kameez suits, dupatta pinned at the shoulders, her long curly hair tied in a thick braid, Namitha Vigneshwar was the epitome of the *nalla ponnu* (good girl) typology as understood on campus: she worked hard to secure good marks, did not explicitly break hostel rules, did not have a boyfriend, and was also fun-loving and popular in her class. Her family lived in Erode, forty kilometres west of Salem, and she went home every weekend to see her parents and younger brother. The 'freedom' to travel home had been granted by her parents only that year; in the years before, her parents had made it clear that she was not to leave the hostel and take the State Transport Corporation bus to come home. Instead, her parents would try to drive down to see her every weekend. The rules of the college hostel also aided them in enforcing this; the college did not allow her to make any visits without their written permission. However, two years of uneventfulness had convinced her parents that she could travel home by herself.

Namitha's case is not an isolated one in Tamil Nadu where rules and regulations that qualify women's presence in spaces of higher education and employment are commonplace, and young women do not often have a chance to complete their higher education or earn their living even in dire circumstances. Yet Namitha's father had gone against many social and community norms and made higher education for his daughter a priority because the family had gone through hard times, and he was convinced that education was a must in order to secure her future, given his own shaky past. There was a time when Namitha's parents could not even afford to pay the fees for her primary education; she could recall being punished for not paying her tuition fees of a few hundred rupees, when she was enrolled in a government-aided convent school. Fortunately, her parents had successfully overcome their hardships and now ran a successful yarn-spinning mill of their own, housed in a unit behind their home (she referred to it as her father's business, although it was registered under her mother's name). Once the business started doing well, Namitha and her siblings were immediately enrolled in an elite 'super school'[1], where the extracurricular activities included equestrian sports and a single month's fee added up to ₹40,000 (about $485 US). She became fluent in English and scored well in her board exams. But her father had not waited to see what would happen at the state-level admissions for Government Quota engineering seats and had got her directly admitted in CCT, which had a good reputation in the state, by paying a Management Fee of ₹3 lakhs (around $3650 US).

When I met Namitha during her third year at CCT, she was standing at the threshold of the 'campus placement' season and had to decide her course of action for the future. By her own assessment, continuing her studies by enrolling in an MBA or MTech programme would be more acceptable to the community elders who had already cautioned her father against sending her to an engineering college. In her mother's opinion also, this was the right path for her: enrolment in higher education meant she would remain dependent on her father and he would not have to suffer the infamy of supplementing the household income with his daughter's earnings. Moreover, higher education might prove to be more 'respectable' than a job in the IT industry, which would complicate her marriage prospects. Once she started working in the IT industry, her parents would have to find her a groom of higher qualifications and salary to achieve a balance in status. Moreover, if she took an IT job, she would no longer be living at home or in a hostel but probably in rented accommodation or as a paying guest somewhere. Her parents would then be concerned about her lifestyle as a young, single, working woman living alone in a metropolitan city. Having witnessed her parents' toil, however,

---

[1] See a more detailed reference to 'super schools' in Chapter 1 in this volume and Elangovan (2012).

Namitha wanted to do something for her family by getting a job and supplementing the family income. 'But it all depends upon my father's decision' ('*Paakkanum, appa enna sollraangalo, adhai porthu dhaan*'). She continued to work hard on mock tests and practising other skills that would be needed while appearing for campus placements, saying she needed to be prepared in case her father agreed to let her enter paid employment.

'Do you know why my parents named me Namitha?', she asked me once. The previous night, I had witnessed Namitha's friends teasing her that she had been named after a South Indian star who was known for her well-endowed figure. 'Look at her', the young women had said, making comments about her physique. '... That's why she is named Namitha!', they had said, giggling. Although Namitha had laughed off her friends' comments that evening with mock anger, they had obviously riled her. On finding a moment alone with me, she could not resist the urge to clarify the matter of her namesake. 'I know for sure that you have not been named after the actor Namitha', I said to reassure her. Namitha continued, 'My father named me after an aunty who has done very well in life. They are well-to-do ... rich ... they have a big house in Chennai. My father decided to name me after her so that I can have a role model, I should be like that ... "*Irundha ... andha maadri irukanum*" he had said'

Young Women in Education

Women's unprecedented entry into higher education—nearly forty-eight per cent of students in higher education in 2017–2018 were women (John 2019) with a significant number of these women in STEM disciplines[2]— has opened various respectable job opportunities to women that are increasingly acceptable to parents and families. Women's increased entry into streams such as engineering in no small measure has played a contributory role in driving the demand for engineering colleges. In fact, women's presence on an almost equal footing with men (the

---

[2] India ranks second in the world for countries with population above ten million for women in STEM (World Bank Gender Data Portal 2018).

ratio of male to female was 0.41 in 2017, according to AICTE website (Enrolments: Gender and Category Wise) is often celebrated as the absence of factors such as the 'leaky pipeline' and 'chilly climate' which have been used to describe science-based educational milieus in the West (Seymour and Hewitt 1997; Jacob Blickenstaff 2005; Brainard and Carlin 2013; Walker 2001).

Yet, Namitha's narrative is emblematic of how women find themselves enrolled in engineering courses through a combination of factors, 'a curious intermingling' (Subrahmanyan 1998, 19) of social structures, norms, and decisions by figures in authority rather than individual preference. Gender figures very significantly in Namitha's narrative: she consistently alludes to men's roles as patriarchs, breadwinners, and decision-makers, while framing her own gender and position in the life cycle as being limitations to entering formal employment and the workplace. Though her family's life has been shaped by aspiration and mobility, she is unsure whether she herself will be able to signal her aspirations and desires of employment in the IT sector. Namitha has a clear understanding of what she can and cannot do: not being able to take local transport to go home versus her parents coming to see her, getting into employment versus pursuing further studies; embodying respectability while distancing herself from the sexual; and hemming in her own aspirations based on her father's decisions. While Namitha's father presents a portrait of paternal aspiration, a man who is willing to shift the boundaries of what is acceptable in the community to ensure a successful IT career for his daughter,[3] she also presents him as an exception in a milieu marked by gendered roles and division of labour, even though the separation between these roles often blur in reality. Men *toil* and *suffer* in the workplace, make decisions for the family, while women care for the children, cook for the family; they are second in command. When men cannot fulfil their roles as primary breadwinners, women engage in the formal economy, although this is often uncomfortable for the men and the household, even resulting in shame (*vekkam*) as it goes against the gendered spatio-cultural code that posits men and women as being part of the

---

[3] See Alice Clark (2016) on the wider phenomenon of status raising as a 'father–daughter project' in neoliberal India.

exterior and interior respectively.⁴ A daughter's education is important, but so is her marriage; the perception among families is that her 'purity' must be guarded—a college must therefore reproduce the protectionist paradigm of the home and 'manage' her sexual awakening. The 'new' workspaces to which women have access, as a result of their education, could be morally corrupt, even as they offer lucrative opportunities to consolidate a middle-class identity.

However, even as she demonstrates being embedded in this discourse, it is not as if Namitha lacks agency—what is striking is the contingent strategies up her sleeve in thinking of what is possible as a future course of action to secure her family's future, given the possible limitations set by her parents, extended family, and caste elders. Even as she reiterates gendered discourses, Namitha was able to carve an identity for herself as a 'good girl' (*nalla ponnu*) who has not given her parents grief, or caused social embarrassment by her behaviour, and has gradually opened up opportunities for herself. Namitha's agency is to be viewed, as Kalpana Ram (2013, 260) suggests, in reference to her study of Tamil women as 'an intricate combination of the conscious and the habitual, where freedom relies on the integration of the past with the habitual'.

CCT had done nothing much to change this essentially patriarchal scaffolding of Namitha's life, which marked men and women as essentially 'different': disciplines, tasks, streams, and spaces were explicitly and implicitly coded as 'masculine' or 'feminine'. The institution reinforced these meanings and boundaries in multiple ways. The college dress code⁵ prescribes different dress codes for men and women— women can only wear sleeved salwar kameezes, the slit of the kameez starting only ten centimetres (four inches) below the hip. Men can only wear collared t-shirts or shirts, with trousers or jeans.⁶ Despite the open and inviting architecture of the college, many corridors are designated as

---

⁴ Reflecting this, the Tamil word for husband is *purusan* (of the *puram*: of the exterior) and wife is *manaivi* (stemming from *manai*—home; the interior).

⁵ The college dress code is displayed prominently at the entrance of the main administrative block of the college on a board that is about 1.2 metres (four feet) long and 0.6 metres (two feet) wide. The dress code is for everyday wear. On formal occasions, students wear the college uniform—a navy blue trouser suit for men and women. I analyse this uniform in Chapter 3 in this volume.

⁶ The dress code was followed in spirit if not in letter. The men did wear black shirts or strategized to have messages printed on 'Department T-shirts'. Women occasionally wore tight-fitting leggings.

out of bounds for students: for young women, presence in the recessed spaces and balconies in the women's hostel that overlook the men's hostel can earn a warning from the security guard downstairs who is keeping a vigilant eye. At hostel orientation meetings, gendered boundaries—both spatial and temporal—are reinforced. Moreover, each building's student-to-teacher ratio is carefully proportioned to ensure that teachers can keep an eye on students. One of the teachers stated that such a ratio ensured that they can monitor friendships and cross-sex interaction closely. When teachers felt that a young man and woman were spending too much time with one another, they would be called in, and students would be 'counselled' against the relationship. The teacher recalled an incident in which she had reprimanded a couple for spending too much time together, and the male student had protested saying he was only chatting with his *akka ponnu* (sister's daughter), to whom he was engaged to be married. However, the teacher had told them there would be no exceptions and they cannot be seen spending time exclusively with one another on the college premises. Thus, even when parents tend to look at relationships as desirable or legitimate, the college tends to take a stand against intimacy between male and female students. 'They should not misuse their freedom', 'hygiene in interactions must be maintained', the teachers and management enphasized at several interactions. Therefore, even though there is no explicit rule against men and women interacting with each other (which makes CCT a 'cool' college, according to my interlocutors), security guards are deployed in the canteen and at various social and cultural events to ensure that students do not sit together or mingle. Usage of mobile phones is perceived to reflect a students' 'character', and often become objects of contention between wardens and students.

Spaces such as playing fields also remain gender segregated—while hundreds of men descend upon the ground in the evenings to play cricket and football, there was not a single woman to be seen, except on occasion, the college's official team in the company of a coach. Young women, who did want to play, do so in the sheltered courtyards of the hostel premises, away from the public eye. Young men and women engage in petty rivalries in the classroom space, some of which explicitly focused on gender. It was not uncommon for young men to organize gendered factions during elections for positions such as 'Department

Secretary' and 'Assistant Department Secretary',[7] or say that they could not stand a woman's authority or take orders from her ('*Naan eppavum oru ponnukku keezha vaela seyyamaataen!*', 'I will definitely never work under a woman!').

I do not mean to paint all young men with the same brush: men also grapple with different subjectivities within the system of patriarchy. It exerts its own pressures on men, chief among which is the necessity to emerge as a successful breadwinner in a milieu of financial precariousness. Many dimensions of masculinity are also on display: men working with machines, enjoying sports, standing at the windows watching passersby, sometimes catcalling or embarrassing them. Their gaze may often be on young women, but they are also objects of surveillance.

Given the location of the engineering college, engineering disciplines also took on gendered hues: streams such as the Mechanical Engineering continue to be male-dominated, although young women now dominate disciplines associated with ICT such as Computer Science (CSE), Electronics and Communication Engineering (ECE), and Information Technology (IT). This can be seen as a positive effect of the changes in the post-liberalisation period that has included globalization, consumerism, as well as cinematic and state espousal of the feminine presence in the IT workplace as a key part of modernity (Pal 2019). As a result, women are more visible than ever before in professional education and the globalized workplace of Information Technology (Chanana 2001; Vera-Sanso 2006; Radhakrishnan 2011; Belliappa 2013). The numerical dominance and positive portrayals on screen, however, does not mean a favourable disposition to women's increased presence in the workplace. Though air-conditioned workspaces and desk-based jobs may appear more suitable for women than the factory floor and construction sites, middle-class families in Tamil Nadu feel very ambiguously about women in the IT workplace and the globalized lifestyles of IT professionals. These discourses mesh with older discourses about women's propriety in the

---

[7] These positions were filled by elected representatives from the fourth year and third year respectively. The college followed a policy of alternating gender in who could contest the elections each year. Thus, in 2014–2015, if the post of Department Secretary was occupied by a male, in the following year the position would be open to only female candidates. The Mechanical stream, however, had only male representatives so far, owing to numerical dominance.

workplace (Vera-Sanso 2006) as well as moral discourses about the 'IT class' in south India (Gilbertson 2018). Therefore, increased presence of women in engineering colleges has more to do with families' ideas of education for daughters as a value in itself—a sign of their middle-class status and their 'worldliness' (De Neve 2011, 84; Gilbertson 2018, 138–139).

Yet as Namitha's narrative establishes, decisions to enter higher education do lead to employment and notions of 'respectability' are negotiated and renewed, even though a moral conflict remains between cultural notions that women should stay at home and economic realities that require a double income. Such dilemmas are not just about femininity or patriarchal rules for women, but reinforce the twentieth-century 'hegemonic ideal' of a family form 'at whose heart (and head) stands the man of substance:the man with financial resources, earning power, a network of dependents of whom a wife and children are most crucial (Osella and Osella 2006, 2–3; Abraham 2011). This ideal has continued to retain a grip over middle-class imagination even as roles of men and women in marriage and family undergo change, especially in upper middle-class families where women work not out of necessity.

With women's education and work placed within such a schema, it seems only natural that the entire model of private higher education is designed to enhance respectability and prestige of the patriarchal family. Such colleges have emerged as feasible options for women to enter science and engineering careers while maintaining respectability. Thus, in CCT, like in many other institutions, various technologies are used to regulate young people's behaviour to reinforce and strengthen a gendered *habitus*, in addition to producing the student as a disciplined individual: one who is on time, one who attends college every day, one who studies after dinner, one who is a good child to his/her parents, one who uses technology only in academically productive ways, and one who fulfils the middle-class expectation of maintaining respectability while on campus. Such technologies meant to produce 'compliant subjects' imbricate with gendered discourses of safety and sexuality in multiple ways right from the time of admission to the dynamics of everyday college life, crucially forging meanings about education and what young women can do with that education.

Gendering Enrolment and Disciplines

Gender is strongly reflected in the enrolments in CCT as seen in Figure 4.1. CCT has a total of seven departments: Civil Engineering (Civil), Computer Science and Engineering (CSE), Electronics and Communication Engineering (ECE), Electrical and Electronics Engineering (EEE), Fashion/Textile Technology (TT/FT), Information Technology (IT), and Mechanical Engineering (MEC). Among these branches, training in CSE, ECE, and IT is considered most suitable for software jobs as they deal with the intricacies of circuits, content writing, and coding. In CSE, ECE, and IT, across the four batches I surveyed, the number of female students outnumbered the male students. Many young women I interviewed in these departments said that their parents had explicitly stated a preference for these streams because the labour did not involve working on a shop floor or a construction site, but jobs in the software industry—in an air-conditioned office, at a desk in front of a computer, with their daily commute possibly taken care of by the company.[8] In fact, many students said that even though they had secured a place in another college through the counselling process, their parents had forced them to decline the place and sought admission through the management quota in this college, close to home, even if it involved high capitation fees. Many women interlocutors pointed out they had been enrolled in a provincial college like CCT only because it was close to their homes, when their brothers had been sent to Chennai, the metropolis, to make their way in the world, often in departments of their interest. Another interesting connotation that had become attached to the IT-oriented streams was that it was now academically 'inferior', associated with the

---

[8] In Smitha Radhakrishnan's (2011) account of gender in the IT industry, 'content writing' and 'graphics' are considered female domains, and 'coding' emerges as the masculine subject. Even though women may enter coding jobs, they are usually involved with quality and maintenance whereas research and development work emerges as 'the most male segment of the workforce, where deadlines are stiff, pressure is high, and the hours unpredictable' (2011, 214–215). Reflecting the above ideological construction, women working in the private sector were observed performing a particular kind of idealized 'Indian' femininity: 'a femininity that reflects the back-and-forth cultural processes that women experience and navigate, from home to work and back again' (Radhakrishnan 2011, 194). Radhakrishnan finds that such performances include the observation of cultural events such as Sari Day and 'choosing' tasks that were coded as feminine in the work arena. Conflicts, tensions, and uncertainties did not find expression in the workplace, but were relegated to be dealt with by what the IT professionals called 'feminine professionalism' (2011, 194).

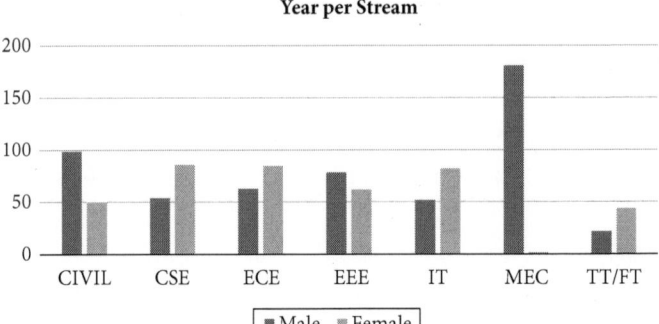

Figure 4.1 Average number of male and female students per year in various streams in 2014.
Source: Compiled by author from CCT Records 2013–2014 that was shared by the management.

lowest cut-offs and 'lack of merit' because of the sheer number of departments offering such courses. CCT itself had three departments associated with the ICT—a trend in other colleges as well. The law of demand and supply ensured that 'anyone could get into an IT-oriented department', if they wanted.

Reflecting this, other fields such as Mechanical and Civil Engineering, which involve careers on the factory floor or on a construction site remain male-dominated: in the cohort of 2012, there were four young women in a Mechanical class of 228 students. Having four female students enrolled in a single class in the Mechanical Department was unprecedented in the history of the department (established in 1997). The classes of 2008 and 2009 had a single female student each. When I interviewed the four young women about their experiences on the day of admission, Jothi narrated how she had originally declined the place when she had found that there were no other female students enrolled—the thought of being the only woman in a classroom and one of few women in the department made her apprehensive.

On the contrary, Monika had decided to take up the seat no matter what. Her choice of stream had been motivated by the desire to improve her father's factory, a unit that fabricated metal components such as grilles, railings, and mesh. Her father's dream was that she would take over the unit and turn it into something large scale, she said. All her cousins had also opted to study Mechanical Engineering, and she would not be the only exception in the family. Though she was from provincial Andhra Pradesh, she had completed her schooling in Hyderabad at an IIT entrance coaching centre that had received accreditation as a pre-university institution. She had not managed to make the cut-offs required by IITs, but got admission into a second-rung institute, a state-run technological university. However, when she and her father had visited the campus to pay the fees and complete the enrolment procedures, violent protests for enforcing caste-based reservations had broken out and left her father unconvinced about the suitability of the college for her. He decided to change tack, and on the recommendation of a relative living in Tamil Nadu, sought admission at CCT by paying a management fee. A disappointed Monika felt she could not further compromise on her higher studies; she wanted her choice of field even if she did not get her choice of college.

During the interview process, Monika said she was relentlessly quizzed about whether she would be able to cope with the skewed gender ratio in class, but was firm about her decision. Once Monika enrolled, a member of the college administration had called Jothi to tell her that she would not be the only female student if she enrolled. A relieved Jothi had come back and signed up for Mechanical Engineering. The other young woman, Bhuvana, who came from a reserved category, was concerned about the number of young women in the Mechanical stream, but had no way of knowing because she came through the Single Window System counselling process for government quota seats, during which details of the gender breakdown in the classroom were not shared. She opted for CCT because it was closer to her residence. She had 'no choice' in the matter, she said.

In the Civil Engineering stream, men outnumbered women by a large margin, but the percentage of young women was higher when compared to the Mechanical stream. In some batches, the number of young women was as high as fifty out of 127. But that is perhaps an exception: going by the number of enrolments over four years, the young women average about a third of the class. For many in this branch, aspiring to be a civil engineer is a conscious decision to stay away from the post-liberalization IT boom, with its morally suspect consumerist lifestyle, rooting themselves instead in an older conception of post-independence Nehruvian discourse of development and nationalism.[9] Many of the students in the Civil Engineering stream were also from countries such as Nepal and Bhutan, who said that since their countries were still 'developing', their contribution as Civil Engineers would be valued in building the nation.

On the other hand, the Fashion/Textile Technology stream was considered more appropriate for women students to join. However, the sex ratio in the course did not show as skewed an imbalance as the Mechanical batches. Except for one batch that had seven male students when compared to forty-four female students, the other batches had comparable numbers. In fact, the number of male enrolments tends to increase after the first year because students with a vocational training diploma in the same stream were allowed to skip the first year and join the second year ('lateral entry'). Also, in the Kongunadu region of Tamil Nadu, textile is a traditional source of livelihood. Apart from the large-scale textile mills in the region, there are also smaller spinning, weaving, knitting, and dying units, and textile is not considered exclusively women's labour but a respectable source of income for many families (De Neve 2004; 2011). In fact, it is often the spaces of textile work that men's masculine subjectivities are formed—whether as patron, 'big man', householder, or as a cosmopolitan man full of 'style' (De Neve 2004, 60–95). In fact, many families enrolled their sons and daughters to the department in order to gain a professional degree in textile before joining the family business.

The Fashion/Textile Technology stream has also acquired a change in 'status': from having 'working class' connotations at one point of time, especially given the lateral entry of students from vocational training programmes, to having the status of what they called a 'premium programme'. For the academic year 2014–2015, the places allotted as Management Quota in the FT/TT batch had been filled much before the class twelve results were out. Interlocutors in the class rationalized their choice either through 'family tradition' or 'scope', 'This is what my father and grandfather did ... and I believe this work is cut out for me', said Bhalakumaran, describing his family occupation as 'weaving'. Another student Sankaran

---

[9] I have already discussed this in detail in Chapter 3 in this volume. Such a discourse was widely prevalent in their department events.

said that when he had first joined FT/TT, neighbours and relatives asked him if he was going to be a mere machine operator in a garment factory, but he had told them that 'actual skill' was involved—it was a 'real' engineering degree. They said that the recruitment process had been easier on students from FT/TT, as there were fewer colleges that offered FT, and they were in the 'textile belt'. Students often said, with a touch of pride, that even students from more traditional engineering disciplines such as Civil Engineering appear for placements in textile companies.

Another student Thamarai said that she was not interested in joining engineering 'chumma ... simply'; she wanted to study something that was not the usual Computer Science or IT, and FT gave her the option of doing something else, 'something different'. For many students, the choice of FT emerges as a desirable option because it had fewer Mathematics papers than other streams. Thus, it offered an opportunity for students to acquire the 'status' of having an engineering degree, and acquire 'exposure'— master soft skills, improve English—without having to deal with 'difficult technical subjects'. Such 'exposure' would become highly valuable in family businesses, even if they did not eventually go 'out' to work (also see De Neve 2011). In some colleges, such an agenda of being at engineering college to acquire other social skills had been taken to another level, turning the college into a kind of 'finishing school': this was visible at progammes such as after-class beauty workshops offered in some women's engineering colleges.

The construction of certain streams as 'masculine', 'feminine', or 'lower class' is not produced by parents or by families alone. Such discourses also carry traction within the college, among teachers, students, administrators, and members of the college management. As Monika recounted, her admission into the Mechanical stream at CCT had to withstand the 'counselling' she received during admission. When Jothi had received similar 'counselling', her response had been to say, '*kandipaan mudiyaadhu*' ('definitely not'): she would never be able to overcome the fact that she would be the only female in a class of a hundred-odd men. She would be intimidated, lonely, and vulnerable to *vekkam* (shyness) and risk. Her parents also did not think that it would be a good idea. Such conversations reveal the 'institutional validation' given to parents' anxieties and fears.

The following is a short excerpt from my field notes on how such institutional validation is produced. The conversation that I reproduce here is useful in accounting for the subtle semantic shifts and manipulation that reveal the positions of men and women in the management and administration as they speak about gender identities and gendered choices.

Salem
7 July, 2014

I went in search of the Management Information Systems (MIS) division in CCT today in order to collect quantitative data for my study. It was a large airy ground floor room that overlooked a patch of greenery; various desktop computers were set around the room and everyone in the room was absorbed in their work. Interestingly, even though I had gone in search of quantitative data that afternoon, what emerged was an interesting analysis by two teachers about the processes of gendering in engineering colleges.

After I introduced myself to the person heading the MIS team, Dr P., I explained the purpose of my visit and showed her the permission letter from the principal. She seemed satisfied with my credentials and asked me to sit down for a chat. Dr P. said that she herself holds a PhD in Artificial Intelligence and introduced her male colleague, Dr S, as a cryptologist.

I started reeling off the information I needed (break up of gender and community categories and the number of first graduates in each class), and she said there should be no problem. MIS stored all such data in one master file on Microsoft Excel; whatever information I wanted could easily be selected from the master file and printed out. I thanked her and asked her if she has been able to decipher any dominant trends in the data. Looking at the paper on which she had jotted down my requirements, she started by telling me that the number of males to female students is comparable (fifty-seven per cent of the student body is male). I asked her if she knew why, and she said that it was primarily because parents want to keep their daughters close to home and send their sons to study in metropolitan centres such as Chennai. In colleges in Chennai, there are significantly more males than females in the colleges, she said. Again, when I asked her why, pat came the reply: 'to keep them under their control', though she quickly altered her response, 'parents are very attached to their daughters, and would want to keep them in the home, feeding them well, and looking after their health. As soon as they are done studying, they are married, and then God knows where they will be', she trailed off, 'maybe abroad'.

I could not help thinking about what I had read about 'this region': Salem district, had the dubious distinction of having the worst sex ratio among all districts in all of India in the 1980s and 1990s: 849 females for every 1,000 males (1991 Census; Chunkath and Athreya 1997). The district's skewed sex ratio results from the social legitimacy given to the female infanticide by the dominant caste groups in the 'female infanticide belt' running from Madurai and Theni districts in the south along the west through Dindigul, Karur, Namakkal, Salem, Dharmapuri districts to Vellore in the north. I nodded my head instead, and she continued to tell me that the 'meritocratic bent' in CCT ensures that young women who do better in class twelve exams come to the college. In this, she was interrupted midway by Dr S. who says that even though the pass percentage is better among young women, the individual toppers are usually men. Dr S. also said that more women also join 'soft' and 'easy' streams such as information technology, computer science, and electronics and communication. Parents also force their children to take up such streams, he said, and then corrected himself, 'they don't force; they push for a certain choice'.

The conversation between Dr P. and Dr S. serves as an illuminating account of how certain disciplines are constructed, and meanings debated through the processes of discursive construction, particularly in gendering various subjects, streams, and notions of success. Thus, women's academic expertise or excellence in engineering college was not called in question. This is consistent with research that men are not thought to be better than women at mathematics, science, and technology in the Indian context (Mukhopadhyay 2004, 476, 481; Subrahmanyan 1998, 90), though Dr S. seemed to be hinting at a difference in performance by suggesting that individual toppers are usually men even though the pass percentage is better among women. However, Dr S.'s pun on 'soft' indicates an inherent gendering of certain engineering branches as more suitable for women, while streams such as 'Civil' and 'Mechanical' are seen as indicative of 'hard' labour. In doing so, Dr S. crafts a gendered relationship to certain kinds of technology and workspaces through language. This kind of Butlerian construction reveals the mutual constitution of technology and gender in making streams 'hard' or 'soft'. These kinds of gendered presumptions are, of course, not exceptional, as revealed in the enrolment numbers in CCT and shared by the middle class across the country, and even in the developed world (Walker 2001; Seymour 1995). For students such as Monika, Jothi, and Bhuvana, who had gone against the grain, such discourses seemed superficial, and they held on to their interests convinced that they could excel in their chosen field. On days that we had to visit the lab/workshop, the young women appeared enthusiastic about working with the lathe and welding machines, eagerly going forward when the teacher-in-charge asked for volunteers to help demonstrate exercises. When I asked them whether they enjoyed time in the workshop, Bhuvana said 'I love [doing] this, this is why we joined Mechanical' ('*Idhu romba pidikkum, adhanaala thaan naan Mechanical join pannaen*'). Even though there were gendered allusions to mastery of technology such as the Department T-shirt, which read 'When we screw, even metals cry' (see Figure 4.2), such interactions did not seem to be present in the workshop or at least did not happen in my presence (cf. Cross 2012).

Kiruba Lakshmi, a student from the ME (Master of Engineering) Mechanical stream—the sole female student in her entire class—however, said that work on the shop floor such as welding and filing could get very difficult for women. Moreover, 'it is intimidating to imagine being the sole woman in a shop floor full of men, with a male supervisor. What would their relationship be like? How will they control the woman?' she said. Yet, she said that there are many advantages for women in Mechanical Engineering:

Mechanical Engineers get government jobs ... since there is a quota for women, very often it is the women who get the jobs. If there are 1 lakh (1,00,000) men and 1 (1,00,000) lakh women candidates, women will get the job first. Knowing this, women take Mechanical (which doesn't have many women) and use it as a strategy to get government employment. Also, many women are here only because their fathers already own industry, and it will be good for them to be a partner/proprietor.

Thus, different engineering streams not only carry different social connotations, but are also rationalized through aspiration, 'passion' for the subject, family backgrounds, family pressures, and strategies for

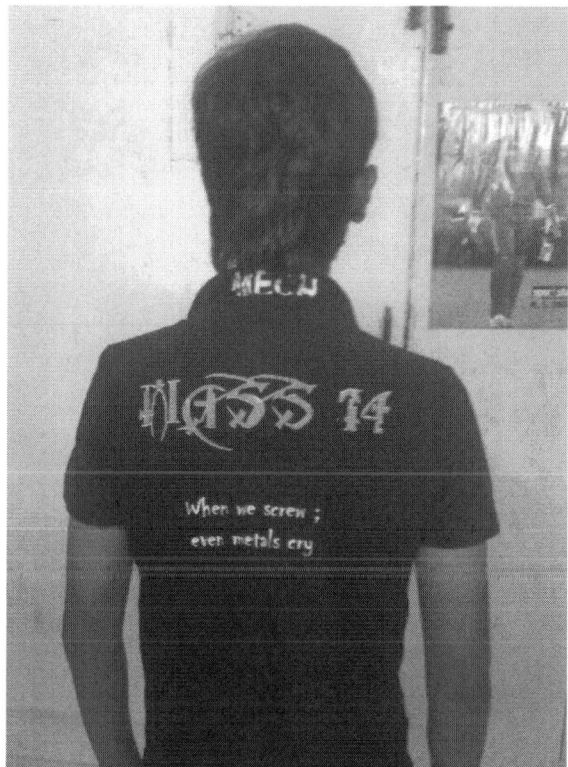

Figure 4.2 The T-shirt designed by students of the Mechanical Engineering Class for their department festival.
Source: Photograph by author (2014).

employment. This shows the multiplicity of ways in which young men and women strategize 'transition' to gendered roles, statuses, and positions, using their engineering degree. However, nowhere are meanings about engineering discipline or its scope, or gender, most powerfully shaped, than in the classroom.

Gender in the Classroom

Even though formally teachers echo a bit of advice from one of the college publications that 'business etiquette is genderless' and that 'men and women are equals in the workplace', gendered ideologies are reproduced in the everyday classroom context. The advice in textbooks which is meant to prepare students for jobs in multinational corporations, however, regularly informs (and disrupts) thinking about what would be respectable and appropriate within a local context. Thus, seating in classrooms is gender-segregated though these are temporarily suspended in workshops and 'personality development' sessions because students have to 'be prepared' to work in teams with male and female members. 'It is quite possible', one faculty member told his class, 'that your boss could be a woman'.

Such sentiments set the tone for rivalries in the classrooms as well: while some were characterized by joking relationships, other described the gendered rivalries in their class as outright *sandai* (fights). There were batches in which men and women were constantly at loggerheads, and the resulting environment was so toxic that the teaching staff had been drawn into brokering peace between the men and women.

Though such heated animosity was not always the case, in an extended case study, I had the opportunity to observe a group of young male and female students working on a vehicle design they intended to take to a competition in Punjab. Two young women and eight boys were part of the project team. Within this group, Monika and Jothi, and Kannan, were close friends and together on a WhatsApp group called 'Three's Company' in which they shared jokes, teased each other, and swapped news, and gossip. Initially, the way the two young women spoke about their close friendship with Kannan, I was convinced that the Internet and the mobile phone presented a channel for egalitarian friendship, away from the intense scrutiny that cross-sex interactions usually invited. However, the young women said that, despite their close bond, their inclusion in the team had come about only as a kind of tokenism because the faculty in charge had insisted that the young men include some women in their group. Yet the young women were excited, and research and development of the design began immediately. In a month's time (which meant that it would fall in the middle of the summer holidays), there would be a regional round in Chennai where they would have to showcase their blueprint. If the team qualified during this round, they would have to implement the design, build a vehicle from scratch, and get it running, so they could present it during the final round in Chandigarh, six months later.

When I met the team the next semester, they were thrilled to tell me that they had cracked the regional round in Chennai. A bit later, when the young men left, Monika and Jothi burst out that they had been cheated out of the opportunity to represent their college at the regional round. They would not have come to know about it if their teacher had not enquired about their pilgrimage to the temple town of Tirupati. Bewildered, the women had said they had not gone on any such trip. That is when the entire story had come tumbling out: the teacher in charge had suggested that the young women present the design in Chennai (since they may not get the opportunity to go to Chandigarh). But, Kannan and the other male team members had told the teacher that the girls were away on a pilgrimage and had taken their place in Chennai. Monika was very upset about the incident, and she put it in context of a couple of statements Kannan had made the previous year, 'that you [the young women] have to merely show your faces to secure marks, whereas we boys have to work hard'.

However, over the next few months, Monika and Jothi put this episode behind them, and worked hard on the project—they negotiated for funds from the college to pay for raw material, discussed logistics, and worked on the design. All along, the young women had to contend with the possibility that they might not actually go to Chandigarh for the final presentation. In order to send the young women, the college would have to send a female lecturer as chaperone, and spend more funds on securing decent accommodation; something that would add to expense and effort. It was also unlikely that the young women's parents would allow them to travel, but the young women were happy that they had at least completed the project. Due to these reasons the young women could not accompany the boys to Chandigarh, even though they would be considered as part of the team. However, a few days later, I met a very upset Monika in the hostel:

> You know the boys did not even register us as team mates! First they told us—your identity cards are not here, how can we register you? When they said this, we got permission to skip class and rushed to the IT centre so we could scan a copy of our ID cards and sent. Even after we did all this, what did the boys do? They did not even bother to download and show. I am so upset! You know, ready to cry only. I am not going to get even a certificate of participation for this! What will I put on my

CV?! You know how much I worked, no? Those boys would not even have finished the project if not for me. Every day talking to the Principal for money, all that I only did ... Kannan is like that only! You remember what he told us last time about girls. We have no value for them.

Although Monika did recognize Kannan's behaviour as unfair, the above episode aptly illustrates how male students are able to take advantage of the traditional restrictions on women—and translate it into credit/merit at an academic level. It also shows the continuities in practices of gender between home and college, where restrictions at home translate into limitations in college activities. In fact, Kannan seemed to feel really disadvantaged in the college when he must participate or compete with women for academic success, and feels the need to manipulate the system—even at the risk of losing his friends.

Monika and Jothi's experience exemplifies how various gendered discourses of respectability, and manipulation by individuals based on gendered logics keeps women students out of everyday activities in college life. Despite the teacher's willingness to include their participation, Jothi and Monika had to actively push for it on multiple levels. Not only did their male team mates devalue their contributions, he restrictions of their parents, teachers, and wardens' also circumscribed their movements, and they were not able to travel and participate in the competition. This highlights how students' identities in the college were not limited to those of being engineering students but were also articulated through their subjectivity to gendered discourses and traditional gender roles.

This was not limited to extracurricular activities outside college alone. The four young women in the Mechanical Department did not think they could even ask to go with their class for the Industrial Visit (IV) [an annual excursion for students to a factory or two, combined with a site of tourist/shopping interest (such as malls)]—'*namaala pogallaam mudiyaadhu, kaeka kooda polaam mudiyaadhu*' ('We cannot go, we cannot even ask if we can go'). I could not help asking why they were being excluded but they said they themselves had volunteered not to go because the IV was an aspect of college life which young men enjoyed the most by indulging in 'male activities' such as drinking, dancing, and smoking, and being typical 'Mech Boys'. By entering the class, the young women had already altered the masculinist culture of the class; they didn't want to interfere further. 'We would be despised if we went on the trip', they said.

The gender dynamics in the Mechanical class presents an interesting case study in the way the women experience the 'homosocial enactment' of their male classmates (Kimmel 2001), but also eventually learn to internalize their constraints. As Gerson and Peiss (1985) have argued, homosociality supplies men 'with resources, skills, solidarity and power' (1985, 321), and reinforces the lines between genders and contributes to the 'symbolic power' (Bourdieu 1990) of men. This gains higher significance and reinforces male hegemony in an epoch 'when the dominance of men is more and more questioned ...' (Gerson and Peiss 1985, 321). The preference for homosocial groups is indeed encouraged by the college (which is invested in values of heteronormativity, despite its disapproval for heterosexual interaction), and simultaneously preferred by students who feel threatened or anxious by the presence of women students who outnumber them in some classes, and compete with them for academic and professional success.[10] These anxieties must also be attributed to the crises in finding employment, and dealing with the paucity of cultural and social capital.[11] Vulnerabilities and expectations are centrally implied in these anxieties—for the young men, the expectation was that they would emerge as breadwinners, responsible householders who would be able to provide (Osella and Osella 2006; Abraham 2011).

Thus, despite being enrolled in the department, the female students did not feel part of it in the same way that male students did. They had to not only negotiate for opportunities that their male classmates took for granted, but had to constantly justify their presence with purpose. Access to opportunities and pleasure alike evaded them if it would take them out of the college space. As illustrated by the narratives above, even though

---

[10] In a way, this situation is comparable to the milieu discussed by Prem Chowdhry (2007) in rural Haryana, where young men are not considered fully adult in the domestic realm until they secure a job and get married. Because of the high rates of unemployment, young men resort to forming collective groups and enacting a public masculinity in order to gain respectability. Chowdhry says that by forming such groups 'Men prove their manhood in the eyes of other men not necessarily through demonstration of wealth and other accomplishments, as suggested by Michael Kimmel (2001), but by supporting/participating and implementing the decisions of the dominant caste male leadership who call the shots in the traditional panchayat (the *khap*)'.

[11] Studies of Tamil youth in Chennai, Bangalore, and Madurai (Nisbett 2009; Nakassis 2010) reiterate how young men feel palpably anxious around young women, especially women who show signs of being 'high class' or come across as competing with them for success or status.

they challenged gender roles and the gendered meanings of their chosen fields, dominant relations remain unchanged.

Gendered Discourse and Performance

'Events', as Norman Fairclough suggests, 'arise out of and are ideologically shaped by relations of power and struggles over power' (2010, 135). In CCT, special occasions such as Women's Day, Sports Day, Hostel Day, and Tamil Mandram (Tamil Forum) were celebrated on consecutive days as part of a cultural week in February 2014. For the entire week, a *shamiana* (large tent) was erected over the entire sports ground in order to transform it into a venue in which the whole student body and staff could be accommodated—all events took place at this venue. On one side was a raised platform, a stage, on which formalities were conducted and cultural performances enacted. On either side of the stage were portraits of the founder and his wife, duly garlanded with fresh flowers each day. Chairs were placed facing the stage for the audience to sit. Even though students attended class during the day and assembled at the ground only in the evenings, anticipation for the evening's event marked these days as special and different from the everyday. The excitement was palpable as students discussed what to wear, who the guests were, and what food would be served.

Some of these events were extremely popular, and all the seats were occupied and more jostled at the sidelines to view the proceedings, dance to the music, and sometimes even throw one another up into the air. Security was tightened on those days, and a barrier of chairs was constructed between what was demarcated to be the male and female sections. Hostel wardens also stood by the barrier to ensure that no student, male or female, dared to breach this boundary. Security guards were stationed at various points to ensure no one left the premises or met on the sidelines.

At such events, performances by orchestra and prominent singers from the Tamil film industry would be accompanied by students of both sexes dancing in a segregated manner. It was not unusual for the orchestra to be interrupted midway for a sneak announcement warning the young women to respect boundaries—'*Ponnungal boundary cross panna mudiyaadhu*', 'Do Not Misuse Your Freedom'. After such announcements, even as the young men continued to leap into the air and jump on chairs, women moved towards the side-lines finding it difficult to even tap their feet. When I asked why we were not making more of the evening, one of the young women said—'You don't know how long memories last here'. Such discourses demonstrate the privileges granted to young men and the apprehension on the part of the management that women could easily provoke the men into misbehaving just by the simple act of crossing some imaginary boundary.

It was not as if the women did not enjoy dancing. At events inside the women's hostels, the atmosphere was completely charged and electric. At a dance competition held inside the hostel premises,[12] many of the women wore lungis and shirts onstage, boisterously imitating the suggestive pelvic thrusts and vigorous dancing style that are the mainstay of *dappankuththu*-inspired item numbers in Tamil movies. They stuck out their tongues and shook their hips suggestively, with so much energy and enjoyment to a medley of the top hits of the year, as the audience around them (consisting solely of women) applauded, yelled, and even cat-called. Although all the participants were female, enactment of the feminine characters were conspicuous by their absence. When present, it was usually as a distant figure dancing with her back to the stage, hair covered, swaying her hips as the opening notes of the music sounded, while most of the dancers aimed to be like the 'hero' characters played by actors such as Vijay or Dhanush. These actors were, after all, their 'idols', photographs of whom were often displayed as their profile pictures on social media, and over whose popularity they fought with their friends! However, interludes of comedy were included in the dance medley that brazenly critiqued the dominant masculinity embodied by the stars—thus, as part of the climatic end of the dance, lungis were grabbed, thrust into the mouth, and held clenched between the teeth, or turned over the head (they wore their salwar kameezes underneath). Or, hips were gyrated in an exaggerated fashion while staring lecherously at the sexualized figure in the background, who, when she finally showed her face,

---

[12] In the run up to Hostel Day, various competitions were organized in the hostels in poster-making, mehendi design, pot-breaking, and dance. Students participated enthusiastically in these events, at the end of which loud music was broadcast and residents took to dancing, in an impromptu fashion, in the sheltered space between the hostels as the wardens looked on fondly. The wardens were periodically dragged into the crowd to join the dancing, which they did.

turned out to be an old woman, dupatta clenched between her teeth, biting her nails as if a young, shy girl.[13] Such 'male comedy-dancing'[14] emerged as the de facto dance genre among the hostel residents; successive groups of dancers mimicked the same style, and were greeted with hoots of laughter and thunderous applause. Only the costumes changed, rustic and rural costumes sometimes giving way to more 'professional' costumes such as black ties and white shirts, as if showing men from village working class transforming to fit into urban service-oriented regimes, but who were invariably engaged in similar performances of masculinity.

When the female-only audience joined in the dancing later in the evening, such a pattern continued, even with wardens and tutors-in-charge[15] reproducing similar performances. In appropriating these steps and costumes, dancing seemed to be constructed as an essentially masculine activity, in which women participated only by embodying the masculine. Yet, it also offered a chance for young women to slip out of the 'demure' comportment required of them in everyday life, and parody the play of masculinities around them in a safe place, out of the public eye (also see Sneha Krishnan 2018). In slipping time and again into comedy, a la Vadivelu and Senthil,[16] women also poked fun at the worlds of male youth, their sexuality, putting them in spots of *vekkam* (such as being caught staring lecherously at an old woman). At the same time, a distance was maintained from embodying female desire or sexualized femininities portrayed in the item numbers by actors such as Namitha, who carry heavily sexualized labels.

Other occasions such as the Women's Day celebrations in the college give a sense of what young women at the college make of their own place and reach in the real world. Women's Day, the world over, generally refers

---

[13] Such episodes bear similarity to forms of mimicry and comedy in Malayalam cinema, as discussed by Jenny Rowena (cited in Lukose 2009, 81), but also to the kinds of masculine embodiments of stars by young men described by Osella and Osella (2004, 224–263). However, this account should also be considered as a corrective to the accounts of young men and their movie heroes—to include young women and their forms of star idolizing.

[14] Raheja and Gold (1996) remind us of the potential of joking and teasing in subverting dominant hierarchical orders.

[15] The designation given to lecturers who stayed in the hostel, who had to help out the warden in her daily responsibilities.

[16] Popular comedians in Tamil cinema, who often embody subaltern figures in films and whose jokes, while being funny, also form incisive critiques of society. Their scenes from films are used extensively in the meme subculture, which is popular among engineering students.

to 8 March, a day that began as a socialist holiday celebrating women's suffrage in Russia and later adopted by the Women's Movement and the United Nations as a focal point for the celebration of women's rights. Over time, it has also been criticized as tokenism, a day to reinforce stereotypes, and as a day sabotaged by capitalism for consumerist ends. In CCT, however, Women's Day was celebrated in the second week of February, and not on 8 March. In this way, it appeared quite discombobulated from the global discourse around Women's Day. Students, too, said they were confused—should they wear sari on 8 March, or on the day CCT designated as Women's Day?

Though some women wore saris (while others reserved it for 8 March), the Women's Day celebration was a relatively solemn event attended by a handful of female students and members of the staff. Like at other cultural events and celebrations, the Women's Day programme began with a prayer song, an invitation for the dignitaries to occupy the dais, a welcome address, speeches by the principal, chairman, and chief guest in English, the vote of thanks, and a cultural programme. However, the celebration saw one change in this pattern: the wives (and in one case, mother) of the members of the management sat on the dais in place of their husbands (or son), along with the chief guests. These positions on the dais were meant to be deeply symbolic, ceremonial positions, granted to the women on 'their' day—where they could experience being on a pedestal as they switched places with the men in their lives.

They were joined by two chief guests: one, a clinical psychologist who regularly appeared on a Tamil TV channel for a tele-counselling show; and another, an alumna of the college who had started an entrepreneurship consultancy. Both addressed the gathering on the links between individual, family, and society. The clinical psychologist started with a famous quip (allegedly) by the eighteenth-century English essayist and lexicographer Samuel Johnson, who on being asked who was better—men or women—is said to have replied: 'Which man, which woman? Johnson realized the importance of the individual, and we do not', she said. 'Not everyone gets what they deserve. That is the sad truth. Yet, it is because we pay too much attention to the collective and not the individual that inequalities exist today', she said, referring to measures such as caste-based reservation. She then asked the audience for the names of a few social reformers. In response, 'Periyar', 'Ram Mohun Roy', and 'Ishwar Chandra

Vidyasagar' were some of the names offered by members of the audience. 'No, we do not need social reformers to reform society', she said, 'all we need is for every individual to reform—and build sensitivity'. She ended her speech by alluding to the nature of troubles plaguing the families she counselled, and followed it up by a plea to the young women students not to neglect their families for the sake of career, and stressed the need for them to reaffirm their roles as mothers, wives, and sisters to attain self-realization and contentment. She ended with an anecdote about a little boy who complained to his grandmother that there were ghosts in his room at night, to which his grandmother replied, '*Adhu paiye illa da, adhu unnoda Appa-Amma*' ('*they are not ghosts, child. They are your parents who [are busy at work all day and] come home only at night*').

The other speaker, the entrepreneur, played a video of a woman with no limbs (hands and legs), whom the speaker described as 'completely independent' as she was capable of many tasks such as eating, cooking, and driving. Applause followed the video. As a postscript, the speaker added that the woman was married, and had married a man who did not pity her, and was too busy to even care for her. At this point, the audience broke into even louder applause. She asked all the women present at the programme to conduct a SWOT analysis of themselves (i.e. evaluate Strength, Weakness, Opportunities, and Threats) and learn to convert the negative traits into positive ones. She said that this was the reason she herself had been successful at work and home, and used the opportunity to introduce her own parents and parents-in-law in the audience, who she said had been very supportive of all her decisions, to which there was much applause again.

The discourse generated at such events reiterates the nature of anxieties that surround women's entry into college space and employment. Despite the aspiration to enrol daughters into engineering colleges, women students were constantly reminded of their future roles as wives and mothers. The Women's Day celebrations had not involved a critique of the structures of patriarchy, in tune with the global discourse on women's rights, but reiterated women's roles as daughters, mothers, and wives. In fact, attention was drawn away from patriarchal structures impeding women's success to emphasize that her success depended solely upon her effort (metaphorically suggested in the video about overcoming handicaps). Moreover, the discourse also reinforced the anxiety of women pursuing success in their careers, and what happens to the patriarchal structure of the family as a result. These discourses were important in creating and sustaining notions about the 'rightful' place of women—within the fold of the family—even while pursuing a career. Such discourses also have an 'everyday' presence in interactions amongst students in the college, as it animated everyday interaction, friendship, constructed rivalries, and caused arguments and quarrels.

In the Hostels

When students returned after a month-long summer holiday to begin a fresh academic year in July 2014, there was a buzz that regulations over student conduct would be made stricter that year. It was not clear why exactly these changes were taking place, but there was a lot of conjecture in the form of whispers and rumours—some said that a young couple had been caught in an intimate position during an industrial visit ('IV') from college, others said that a young woman, a student of the college, had gone to stay with a young man during the holiday instead of going home. Some others attributed the tighter regulation to a report of the 'scandalous behaviour' of the CCT hostel warden in a local tabloid. The evening rag had peppered the story with moralistic rhetoric about how CCT could have a conducive atmosphere for study when it did not practise strict segregation between male and female students as in other colleges in Tamil Nadu. There was also talk about a change in the college management: the secretary of the college board had quit and started his own college. It was alleged that he had fed such articles to the local press to damage the reputation of CCT and woo students to his own college. Others said he would never do such a thing. He was a progressive man, who was responsible for the relaxed ethos of the college. On the contrary, it was because he was gone that the college was going to change, they said.

In short, the college was in the throes of a 'moral panic',[17] and the college management had responded with tighter directives for student

---

[17] A term first used by Stanley Cohen [2011 (1972)], the term has been a staple of youth culture studies to discuss how 'a condition, episode, person, or groups of persons emerge to become defined as a threat to societal values and interests'. Most usefully, Cohen underlined the role of the mass media, politicians, and other conservatives in fanning such panic.

conduct in the college. This news, announced in class by the teachers, whispered in the corridors, gossiped over meals, spread a gloom in the college akin to the dark monsoon clouds gathering outside that July. We began to see many signs of change: one day, when I was sitting in class, a few male teachers stormed in and ordered all the students to hold up the identity cards that they were supposed to be wearing around their necks. After a basic check, students were told to wear their identity cards at all times, and asked to be on time and stay in college through the day. I heard that hostel rooms and dormitories had been similarly 'raided' to ensure that students were not staying in their rooms and 'bunking off' class. Many young women even complained that going to the hostel to fetch a book during class hours had become an ordeal since they had to secure an pass from the heads of their departments for the security guard at the gate between the hostel and academic sections to let them through.

My visits to students' hostel rooms in the evenings came to be dominated by talk of these 'raids' and disciplinary actions: one evening, I witnessed an argument between Monika and her room-mates over whether such regulations were actually warranted. A frustrated Monika had composed a letter addressed to the principal after she was denied exit from the college without an 'Outpass' while suffering severe menstrual cramps. The letter began with praise for the reforms, the recognition that it was meant to make them 'more professional', and give them a competitive edge in the industry. 'However', she had continued, 'we also need to enjoy college life', and 'this kind of existence is without enjoyment', she had written.

Her room-mates were aghast when she read out the letter to them. One room-mate pleaded with her not to send the letter or she would be in trouble, but Monika said she already had a plan: she was going to create a false email ID at a cyber café to send the email. No one would ever come to know who sent it. Another room-mate, Rathna told her that she was crazy (*paithiyam*) to send the letter:

Vekkam illaya? Pasanga onnum solladhappo, nee yaen pesara? Nammala baadhikaradhai vida, vidhigal pasangala adhigama baadhikkum, illaya? Avanga vaazkkai murai, nadavadikkaigal, oor sutharadhu ... ellam. Nee oru ponnu. Avangale summa irukarappo, nee eppadi edhirkalaam? Vera college-i vida namma college evvalavo paravalla. Inge sudhadhiram irukku, nallaa nadatharaanga. Idha poi neeye keduthudaade.

Are you shameless? When the boys are saying nothing, why are you saying anything? You know the rules affect the boys ... their way of life ... their movements, loitering, more than it affects us. You are a girl. When the boys are not saying anything, how can you want to say something and protest? Our college is better than most colleges. We have freedom here. In fact, we are spoilt (pampered) here. Do not go and ruin it for yourself.

In Rathna's words of caution to Monika, three things stand out: the gendered nature of socialization within the peer group, the 'difference' drawn between the life-worlds of men and women, and her portrayal of men as having greater cause for protest as 'loitering' is a way of life for men, not women (also see Phadke, Khan, and Ranade 2011). Although Monika did not seem convinced by her room-mates' pleas at that time, she never emailed the letter: perhaps, writing the letter has given Monika a much-needed cathartic release, or perhaps she had decided to heed to her friends' words and resign herself to the rules.

Other students also seemed resigned to the rules, echoing the contents of Monika's letter in saying these measures had been introduced to make them more disciplined and professional. That would be their everyday life, now that they had enrolled here. '*Namakku idhu ellam mudiyaadhu*'—'We cannot do all this', they said. Strictly speaking, the sentence translates into 'we cannot do all this', referring to something that is externally mandated, action that is restricted, or forbidden. However, given that in theory, there were no rules forbidding Monika from writing to the principal, Rathna's injunction reflects not only a constraint that is internalized, but an action that is performative in the Butlerian sense—reiterative and bringing into force what it states (Butler 1993). Moreover, Rathna starts off by asking her friend, '*Vekkam illiya?*' ('Are you shameless?'). Roughly translated, the word '*vekkam*' means shame, but also shyness and embarrassment, similar to the word '*chammal*' in Malayalam, described by Ritty Lukose (2009, 81–84) in her ethnography of a college in south Kerala. However, respectability also means embodying an understanding of what is *mudiyaadhu* (not possible), could possibly cause *vekkam*, and knowing what should be done at a given moment (*enna mudiyarudu theriyunum*). The triangulation produced by these terms emerges not just as an interpretation of the rules imposed by the college management; it emerges as a central mechanism for negotiating college life as 'a technology of the self' (Foucault 1980). These words are discursive, performative, not completely devoid of one's own agency, but also harken to a code that is located in one's culture, in common sense, and *habitus*.

The idea of '*maanam*' or respectability is centrally implicated in the term '*mudiyaadhu*' used by the women. Some of my interlocutors understood respectability in a proverbial fashion as '*ponnu adungunu*' [young women should be demure] or '*ponnu vaayamoodunu*' [young women should keep their mouths shut]. Though these seem like public norms of silence and submission in order to maintain 'honour', it is important to consider its strategic deployment, as Raheja and Gold (1996) suggest in their expositions of women's folk songs in Rajasthan. Thus, in Rathna's caution to Monika to not voice her resistance, which is a wordy version of '*ponnu vaayamoodunu*', we can also glean the meta-message—that if she sends the letter and gets caught, she would simply be inviting trouble, while the institution would never change its stance based on the letter. It would be better to stay quiet and simply go through the motions. Perhaps then we should read Monika's refusal to send the letter as the receptiveness of the meta-message in Rathna's words. As Veena Das (1988) argues, women 'learn to communicate ... by non-verbal gestures, intonation of speech, and reading meta-narratives in ordinary language' (Das 1988, 198).

Over the last few sections, I have illustrated the various ways in which women struggle, resist, contest, negotiate, and live with the masculinized culture of the college, the rules imposed on them, the subjectivities rendered by the rules, and the specific organization of the performativity around the poles of 'masculine' and 'feminine'. In doing so, I attempt to show the ways in which they deal with multiple frames of gendered constraint and restraint in the pursuit of their engineering education and careers, but also fun and leisure. In portraying these gendered subjectivities of an aspiring 'professional' middle class, I do not mean to emphasize victimhood, or suggest that all constraints and restrictions are internalized. Within the spaces of the hostels, within peer groups, young women do discuss the rules, attempting to make sense or resist them. Some rules emerge as guidelines, helping them manufacture respectability and emerge as 'professionals' as discussed earlier, while others appear rigid and difficult to navigate. Students put up with many restrictions, despite their own judgement of its effectiveness, stating that they did not want to create unnecessary trouble; their only agenda is to graduate from the programme. Women also enjoyed leisure within the recesses of the hostel—it afforded the space for enjoyment, chatter, relaxation, and friendship. In fact, life in the hostel often emerges as a 'playground' where women could let their guard down, at least to an extent, and engage in pleasures and pursuits that would have been frowned upon in their own homes.[18]

However, hostel rooms were also fraught sites where the referents of respectability were hotly contested. For instance, covering the bosom with a dupatta was not prescribed by the college dress code but enforced socially as part of the demure and modest dressing. Thus, even when young women did not *have to* wear a salwar kameez compulsorily, like when they stepped out of campus, they wore shrugs[19] over their short kurtas and jeans. It was trendy but also served as a modesty garment. When they did not, it was common to hear her friends berate her saying in Tanglish[20] '*cover your maanam*'. In other words, the young women strived to embody what they deemed as an appropriate femininity, in addition to the external policing on college campuses against clothing that are considered 'modern' (also see Lukose 2009, 89–90). One of my interlocutors summed it up as 'when wearing sari or salwar kameez, girls did not earn as many stares and comments from the boys because the girls looked like their mothers. It earns us respect'.

Similarly, when linked with a young man, the common protest was, '*ennoda maanatha vaangadhe*' ('do not embarrass me'). A young

---

[18] Jenny Garber (1991) suggests that girls often have their own alternative ways of organizing their cultural life, despite their lack of freedom. 'Girls negotiate a different leisure space and different personal spaces than those inhabited by boys. This in turn offers them different possibilities for resistance, if indeed that is the right word to use' they say. Her study of the 'teeny-bopper' subculture (among pre-teen girls in the UK) showed how women gained private space in strong friendship bonds. This helped them 'remain inscrutable' to the outside world of parents, teachers, and boys as well. Garber looks at such friendships as a 'way of buying time' from the real world of sexual encounters, while at the same time imagining these encounters, with the help of the images and commodities supplied by the commercial mainstream, within the 'safe space of the all-female friendship group'. Despite the cultural specificities of the above studies, I found this useful to think through how my female interlocutors organized their own leisure practices.
[19] A close-fitting cardigan or jacket, cut short at the front and back so that it covered only the arms and shoulders, and when buttoned up, the bosom.
[20] A portmanteau term combining English and Tamil, combining words and phrases from both languages—a common slang among youth.

interlocutor once told me, tears welling up in her eyes, that her teacher had sullied her *maanam* by noting her absence in class and smirking, 'she must be roaming around somewhere'[21] when she had actually gone home to visit her parents. 'I don't feel like going to class anymore. That comment has hurt me a lot. What will all my class boys be thinking about me now?'.[22] In the college therefore, respectability lies in cultivating a self, a set of dispositions that avoided being positioned with the sexual, the vulgar, the dirty (*asingam*), and the vociferous. It meant cultivating *maanam* through sartorial choices and a carefully considered code of conduct and everyday interaction that distanced them from the above undesirable traits. The culturally contingent ways in which words such as *maanam* emerge reveals the choices, compulsions, and constraints that frame women's place in higher education—such words and actions are enmeshed at every stage of college life from admission, everyday life, special events, to campus placements, as shown earlier. Similarly, disciplining according to gendered discourse seeped down to every aspect of social life, including the semi-private recesses of the hostel.

The Boundaries of Friendship

CCT had four hostels for women, categorized according to year, creating 'age sets' which drove primary peer group interaction. Because it was semi-private and men were not allowed within its premises, young women were less self-conscious within the hostel premises: they felt free to wear clothes of their choice, play games, celebrate events such as birthdays, dance freely, watch films, gossip about movie stars, and so on.[23] Peer

---

[21] Shilpa Phadke, Sameera Khan, and Shilpa Ranade (2011) unpack the gendered relations to space in their book *Why Loiter*, and reconfigure 'loitering' as a political act. The teacher's remark refers to the gendered and sexualized implications of aimless loitering.

[22] Martyn Rogers (2008) notes that such questions, which hint at sexual immodesty, raised in contexts of 'Eve teasing' or sexual harassment cases have even resulted in suicide. Such hints are therefore taken very seriously in the Tamil Nadu context.

[23] Apart from the expected interest in Tamil cinema, my interlocutors avidly followed Malayalam cinema, Bollywood, English films, British, and American sitcoms. Though not widely popularly, a few young women also watched Korean drama (soap opera) and Japanese animé because of the easy availability of subtitles, even if their grasp over these diverse languages was rudimentary. It is not an unknown fact that cinema is widely consumed in Tamil Nadu (Pandian 1992; Dickey 1993), because of which there is wide social acceptance for watching films as a social activity. Also, mobile and internet technology has translated into better accessibility of films

groups were important because they were the primary circle of support in college. Not only were friendships a source of pleasure and a site of affection and love, they also drove students' leisure pursuits—this could be related to eating, shopping, or visiting spaces such as the theatre, cafes, and malls with the group.[24] They were also the sites of primary care—friends brought food and medicine for each other, cared for each other when sick, going through a heartbreak or family crises, or while suffering menstrual cramps. What was most important about the hostel rooms was also the privacy and intimacy between friends to engage in night-long banter, sharing secrets, 'gossiping', and 'bitching' about their classmates and friends. Within the hostel premises, it was not uncommon to hear endearments such as *chellakutty* ('darling little one') being used to address room-mates and friends, and see other forms of homosocial affection such as hugging, kissing, and parting each other's hair affectionately. These were the freedoms that college life afforded, however circumscribed. They point to the positive aspects of friendship such as camaraderie, resistance, and 'active femininity' produced within the zone of informality. Sneha Krishnan (2014, 69), in her study of women's colleges in Chennai, points to the ways in which cultures in women's hostels exhibit a 'lateral agency'—a momentary pause and sideways movement in which desires, pleasures, and acts, which are rendered invisible in the production of 'ordinary' everyday life, are brought into view. While being in

---

for young women subject to surveillance and excessive control, who may otherwise not have been allowed to visit the cinema hall with their peers. Since Internet technologies enable easy file-sharing and downloading of ('pirated') audio and video files with subtitles, both accessibility and language are not a concern. Moreover, there is a ceiling on the price of cinema tickets in Tamil Nadu—therefore, even spaces such as fancy multiplexes remain fairly accessible to people of all classes.

[24] The private college also plays a role in this, introducing students to certain patterns of consumption and global lifestyles as necessary 'exposure' required to be a future professional, but which would have been out of their reach otherwise. This was individually desired by many students as well. In fact, students often made it a point to visit cosmopolitan malls in Chennai and Coimbatore such as Brookefields Mall as part of the itinerary during college trips (called 'Industrial Visits'). A glittering edifice dedicated to contemporary consumerism in the heart of Coimbatore, Brookefields Mall is located in what was formerly the site of a Brooke Bond tea processing unit. Its changed identity is a veritable indicator of changed aspirations in Kongunadu: middle class aspirations are no longer linked to jobs in such units, rather they are linked to aspirations to visit the site as a consumer with spending power. When students returned from IVs, they would excitedly describe to me the shops and ambience at Brookefields Mall, and share pictures taken against the glitzy interiors. Various studies on the middle classes in India have duly noted these changed aspirations.

hostel, women were outside the adult world where caste and class would prohibit such close contact, and friends were drawn into status-levelling kinship. Close friends would address each other's parents as *amma-appa*, siblings as *akka-thambi*, as if their own—using 'sentiment' as a way of asserting relatedness and friendship (Lambert 2000). As Trawick remarks, such a defiance of the rules of purity was a way of showing and displaying what love was (1992, 150).

Such motifs of egalitarianism have been ubiquitous in sociological literature, especially on friendship of youth (Osella and Osella 1998; Nisbett 2007). However, David Sancho (2012) critiques this literature, saying that too much has been made of egalitarianism, especially in considering practices such as 'treating' as a mode of egalitarianism. He forces a harder look at consumption practices among a group of young boys from an elite school in Kochi, and looks at class assertion and self-making practices that are intertwined with activities such as 'treating' or watching films, which he claims bolstered the 'ego' of his young interlocutors (2012, 135–162). Such activities emerged as a way of asserting hierarchy rather than practising egalitarianism. My time with the young women's peer groups in CCT also showed that emphasis on egalitarianism presents a rather romanticized picture of youth friendships, even though friendships were marked by intimacy, care, and love. In the case of the young women at CCT, consumption emerged as something that united the group as a source of common interest: looking at fashions together online, buying similar bags (that cost about ₹200/$2.99 US), discussing salwar kameez or blouse patterns to instruct the tailor, and strategizing to break the hostel rules to go shopping. Yet, who went out and who stayed in, who engaged in courtship practices such as exchanging flirty text messages with young men, who had a boyfriend and who did not, emerged as frames of inclusion–exclusion in the group.

The gendered ethos of the hostel and importance of respectability as a cornerstone of desirability drove self-making in CCT. This does not mean at all that there was a single mode of thinking about gendered selves. As the ethnography makes clear, gendered evaluations of actions varied across the student body, and different women dealt with risk differently: some wholeheartedly taking huge risks such as strategizing to spend a night out with a boyfriend, others planning to claim access to pleasures linked to consumption, and yet others afraid to step out of the hostel unaccompanied by a figure of authority, depending upon their individual desires and subjectivities. Such interactions often formed the texture of the negotiations in the hostel room as the following excerpt from my field notes shows:

Salem
19 April 2014

After returning from class today, just as I finished sipping the hot sweet tea from the hostel mess, my phone rang. It was Monika. 'Get ready to leave in 10 minutes. Quick. Just be ready. Jeevitha and Haripriya will come to pick you up'. Before I could ask any questions, she disconnected the call. I put down my cup, and started getting ready when there was a knock on the door. It was Jeevitha and Haripriya.

'Akka, seekarama vaanga (come soon).'

Outside, Sowmya and Uma were hanging about, trying to look casual, even as they cast their eyes about restlessly. '(Warden) Ma'am-a kaanom. Seekarama polaam' ('Ma'am is nowhere to be seen, come quickly, let us go'), Uma said.

Jeevitha looked around nervously, 'Enakku romba bayamairukku' ('I am very scared'), as she quickened her pace. While we walked to the gate, the young women explained their plan. They wanted to visit a fair—an 'exhibition'—nearby, but they were afraid the hostel authorities would not give permission. Students were not allowed to go out on weekdays after class. They had divided themselves into two groups to draw less attention to themselves. The second group had already gone ahead.

The young women in our group asked me to walk straight out of the gate and wait, while they handled the guard. A few minutes later, they bounded out, happy with their success, 'Kovilukku poroam', Jeevitha said laughing ('we are going to the temple', referring to the destination stated to the guard).

Reunited with the others, we caught two auto rickshaws and headed to the exhibition grounds. Once there, we bought tickets and headed inside, and spent a happy hour, even though the young women worried from time to time that we had forgotten to bring some kumkumam (vermillion) to lend credibility to the story that we had gone to the temple.

On our way back from the fair, we caught two auto rickshaws to take us back to the college, still excited about our unexpected getaway from the mundane routine on campus. Everyone was discussing their favourite moments, when suddenly Jeevitha gasped, and went silent. When we got out of the rickshaw, she did not attempt to contribute her share of the auto fare, but walked briskly ahead into the college. When we caught up with her, we could see her face was creased with worry. We asked her what happened, and she said that the driver, whose face she had glimpsed in the rear-view mirror, was someone who was well known to her local guardian, a police constable in the city who was her father's colleague. The policeman/local guardian had often called for the auto driver to send parcels of biryani or drop her to the bus stand because her father had made it very clear that she was not to step out of the campus for any reason. And yet, we had unwittingly hailed the same auto driver to bring us back from the fair!

It is not especially remarkable in Tamil Nadu that young women are subject to surveillance. A young woman who wanders alone in the public sphere is extremely vulnerable to gossip, and many young women strategiszd to render themselves as 'conforming' as possible, as Jeevitha's story illustrates. [25] For Jeevitha, this was the usual stance she took when a plan to go somewhere outside the campus was discussed. For instance, when I had just moved to Salem and was still settling in, I had asked Jeevitha to accompany me on a shopping trip to the city. She flatly refused saying '*Enalla vara mudiyaadhu* … I will not be able to come', and that she could not risk being seen wandering about in public places as she had many relatives in and around Salem who would complain to her father that she was loitering about town (*uuru sutrudu*) instead of attending seriously to her studies. When I asked her to at least direct me to a shop in which I could find what I needed, she nervously replied, in a hesitant tone, 'I do not go anywhere, I do not know any of these things. Please do not ask me.'

---

[25] This is also noted by Isabelle Clark-Deċes (2014). Speaking of her twenty-one-year-old research assistant Abi, who had just graduated with a degree in chemical engineering, Clark-Deċes notes that Abi was granted freedom and mobility because she had the responsibility of looking after her family by finding employment. Her frequent trips from her village to Pondicherry (now Puducherry) and back, returning home well after 11 pm were deemed appropriate by her alcoholic father, a tea shop owner, because Abi was looking for a job which would help support the family. Though Abi's job searches were proving to be futile, her perceived 'employability' had cut her enough slack to negotiate with her family on just temporal and spatial boundaries, but marriage as well. Clark-Deċes (2014) attributes it to the sense that her family felt 'left behind', in the context of 'India's new rising middle class'.

Jeevitha's nervousness illustrates how young women's entry into spaces of higher education and employment have been enabled only under conditions of maintaining respectability, and has not legitimatized their entry into public space as such. Though Jeevitha's responses had initially puzzled me, once I got to know her better, I could see from where her anxiety stemmed. Jeevitha described how her father, an officer with the Tamil Nadu Police, had not spoken to her for almost two years, after her class twelve results revealed that she had secured 'only' eighty-eight per cent—not enough to get into the best engineering colleges in Tamil Nadu. He had made it very clear that she had disappointed him, and had her enrolled in what was considered a good college in his native town. Even though he participated in the admissions process and even appointed a local guardian to bring her biryani every week (so that she would not go out in search of a treat), his disappointment had resulted in the complete breakdown of conversation between father and daughter.

Jeevitha described how her move from Chennai, where her parents lived, to Salem had caused a drastic change in her personality. I have '*vekkam*' now, she told me, describing how in Chennai, in her school days, she had been a brash young woman.

> I cannot explain it, Salem has made me shy. Before that, there was not a shy grain in me. But people here ask, why are you wearing modern clothes? Don't you feel shy? How can you wear things like this? How can you go there? How can you be here? Now, I've been made to feel shy (*vekkam*). It has something to do with this place, of being here.

As explained in reference to the vignette about Monika's letter, *vekkam* can be interpreted broadly as shyness and in certain contexts, as shame. It seems that Jeevitha had reinvented herself as a shy and introverted personality to deal with the anxieties inaugurated by her entry into CCT. As a result of such anxiety, one can see in this episode, even an innocuous trip to a fair were laced with fear, of being caught and reproached by the hostel authorities, or a complaint reaching parents. The danger did not lie solely with the wardens and security guards in the hostels reporting them; Jeevitha was afraid of being spotted by relatives and acquaintances of her parents, asking questions about why young women were hanging about with no good reason. Jeevitha's worst fears came true when the rickshaw driver happened to be an acquaintance.

However, it can also be seen that a peer group could display solidarity and collectively strategize an 'escape' from the hostel. As I just described, such strategies were adroitly planned and implemented. Sometimes, even I, as an older research student *akka*, emerged as a 'legitimizing device'—asked to accompany students to certain places in order to lend a veneer of respectability to their escapades. Boundaries therefore appeared emerging as shifting lines—between what is desired, possible, what can be strategized or negotiated, what is simply impossible, and what should be accepted as it is useful to 'become' professional or 'makes sure we do not misuse our freedom'. Strategies are envisioned with great excitement, but also struck down with equal urgency. Responses such as '*mudiyaadhu*' or the remaking of the self through '*vekkam*' show the ways in which women remade their own subjectivities in line with the rules and regulations posed by the administration and their families, even as they accessed spaces of higher education, strategized resistance, conforming, at the same time dreaming of ways to enjoy college life.

However, these decisions were not always easy to make, and called for a certain performativity of respectability irrespective of one's deepest desires. Monika, in private, narrated that as much as she wanted to make greater 'use' of college life, she had to ultimately toe the line. She wanted her friends to accompany her to an amusement park in Coimbatore, for example, but could not keep repeating such pleas to get out and take risks. It would affect her own image in the peer group as someone who courted risk with too much gusto, and could present a 'threat' to the 'respectability' of the peer group. She feared that a radical break would result in exclusion from the group based on moral judgement, and that her friends would 'bitch' (talk negatively) about her and stop talking to her. 'It is more important to have a gang than to go out to enjoy college life', she said. Such performances of respectability were important to retain her membership of the peer group. Such instances show how despite the 'lateral agency' (Krishnan 2015) afforded within one's peer group, respectability needed to be performed even in front of them. Movement away from the 'sexual' and the 'risky' needed to be demonstrated as an important 'technology of the self'—sometimes even emerging as a legitimatizing strategy of staying in higher education. Friendship therefore emerged as a forum of enforcing 'various technologies of the self' such as respectability and showcasing conformism as much as a mechanism of a strategy.

Conclusion

Social constructions and performativity of gender on campus show how young women negotiate college life in conditions where their entry into higher education is premised on increased respectability for their families. I have attempted to establish continuities between life at CCT and the home, in terms of establishing a clear difference between male and female, and the lives of adults and children. I examine how these are produced discursively in the college, whose rules infantilize young adults, and locates young men and women within certain gendered discourses, as it relates to local culture as well as engineering stream. In such discourses, reinforced by figures in authority, women are located within the interior, the familial, reinforcing a hyper-feminine mode of being, even as they study engineering and aspire to become 'employable'. While some of these precepts are outrightly rejected or subverted, many others are also internalized by the students, male and female, and often repeated, contributing to the reproduction of gendered conditions of access to opportunities, pleasures, and pursuits. These discourses span streams of engineering study, inter-collegiate competitions and symposia, and even modes of conduct at events, and in the everyday.

However, rather than leave it at that, I argue that self-making through the trope of respectability is an important mode of negotiating college life, given the shifting boundaries of women's entry into education and paid labour. The performativity of respectability by internalizing and appropriating institutional constraints, and engineering their own boundaries, plays out in every aspect of college life. Students' lives, especially young women's, are a constant negotiation with the rules, small resistances, and ultimately reconciliation to the rules through the trope of respectability. This has also resulted in students always feeling on the edge, unable to feel a sense of safety or belonging, and constantly wracked by anxiety. This seeps down right to the homosocial peer group.

Even though the peer group emerges as a platform of strategy to engage in certain consumerist pleasures such as shopping, it simultaneously forces the manufacture of restraint. Friendship, it emerges, is a process of collaborative effort and solidarity but also defines and determines the course of young women's actions, and the negotiation of college, neighbourhood, city, and locations outside the city. Such mechanisms of control and restraint emerge in different gendered subjectivities and contingent makings and remakings of the self, whether it is Monika's strategy to prune her aspirations for enjoyment in college life, or Jeevitha's steady remaking of self as a person with *vekkam* to be better accepted and respected in the peer group. The making of the self then emerges as a subjective response to whatever is at stake at a particular moment—what Kleinman and Fitz-Henry (2007) call the malleability of 'lived experience'—given the shifting conditions of human life. This is also particularly evident in the delicate domestic arrangements that young people make not only to stay in education, but also in ordering their personal lives in ways that would increase their chances of staying in education, as I elaborate in the next chapter.

5
Negotiating Intimate Risk
Gendered Subjectivities, Performativity, and Self-Making

A key demographic fact of Salem district that remained foremost on my mind while researching for this book was the skewed sex ratio, which had ranked among the lowest in the country in the 1990s. I would often remind myself that the young women around me in CCT had been valued enough to enter higher education, unlike many of their age cohort, 'the missing girls' who had been victims of female infanticide or foeticide (Sen 1990; Patel 2007; Kaur 2016; in Salem context, Chunkath and Athreya 1997). My interlocutors are part of a generation that Alice Clark (2016) calls 'valued daughters', whose lives and education are heavily esteemed by their families, even if they are constrained in other choices. In communities such as the Gounders of Salem and Namakkal districts, this value ascribed to daughters is further complicated by what sociological literature refers to as 'bride shortage': endowing young women with professional qualifications with more negotiation powers when it comes to contracting an alliance. Sharada Srinivasan (2009) in her deeply engaging work on the sex ratio shows that young men who are engaged in traditional occupations such as agriculture without qualifications in higher education are known to struggle to find a bride, whereas women with engineering backgrounds are highly valued.

However, throughout my field notes, I use the phrase 'missing girls' not to refer to victims of femicide, but to young women who went missing from higher education before graduating due to various pressures, mostly from the family. A young woman or two eloped every semester, afraid that they would never be able to convince their parents of a choice match. This meant they would also disappear from college. Many more discontinued their higher education when their families suspected that the young women's presence in the college could disrupt their marriage prospects or the family's respectability. These 'missing girls' speak of the precarious place of young women in higher education—often denied free choice, and having to negotiate between being in a relationship of choice on the one hand, and higher education and employment on the other.

This is a widespread phenomenon all over Tamil Nadu; not limited to rural or 'rurban' regions such as Salem district. Even teachers from colleges in the heart of Chennai said their students went 'missing' for similar reasons. Because anxieties about women's freedom are so entrenched in everyday spaces and discourses in colleges, the phenomenon of 'missing girls' appeared ordinary, hardly problematized or criticized by classmates or college authorities, lending itself to salacious gossip rather than concern. This is evident in the everyday-ness of phrases such as *'veetlai prachnai irrukka?'* ('Is there a problem at home?') as a euphemism to ask if a particular girl has been caught having a boyfriend or has her marriage been fixed suddenly under suspicious circumstances to describe the situations of the 'missing girls'.

In one 'missing girl' case I encountered when at CCT, Sudaroli was compelled to stay home because her family suspected that she was in a relationship because she had turned down an alliance to marry her father's sister's son (*athai paiyyan*), Mahesh. According to her family, Sudaroli had done the unthinkable by refusing a match within such close kin ('a match that one could not say "no" to' [Clark-Decés 2014]), and concluded that she would have done so only because she had a boyfriend, further upsetting her parents. Various family members had added insult to injury by saying that they had noticed that she spent extended periods of time texting and chatting on her mobile phone, sometimes late into the night. 'Whom else could she be chatting with but a boy?', they said. Upset over the aspersions cast on Sudaroli, her father had stopped speaking to her, and had refused his permission for her to return to college. She texted her friends every day about the conundrum she was in, and they discussed the updates again during recess, worry creasing their foreheads. 'Would she ever get back to college now?'.

What was unknown to the family was that Mahesh had first approached Sudaroli with the marriage proposal, even sending her gifts of chocolates and a soft toy.¹ Although she had initially been excited and had described the proposal at length to her friends, even making her best friend chat with Mahesh over the phone,² she had finally refused Mahesh. When discussing her decision to turn him down, she said that she did not want to be 'committed' at such a young age; she wanted to enjoy college life, and did not want to marry Mahesh who tended to get 'possessive' and become suspicious of her outings and the nature of her friendship with other young men. She also had certain misgivings about her father's side of the family—especially the way they had treated her mother—and did not want them to be her in-laws. As if proving her judgement to be right, Mahesh had been hard to shake off. After his personal proposal failed, he had asked his mother to forward a formal proposal for marriage to Sudaroli's parents, despite Sudaroli's rejection putting her in a difficult spot.

Many young women embroiled in such domestic controversies sometimes never made it back. They were made to deregister from the course, kept under 'house arrest', and their marriages were usually fixed soon after. Two weeks later, however, Sudaroli was able to strategize her exit: after giving it much thought, she went up to her father, gave her mobile phone to him saying she did not need it anymore, and sought his permission to go to Salem back to college. Sudaroli's father agreed this time, but she would have to follow some conditions: she would have to return to her *athai's* house within fifteen minutes of the end of class (she lived with another of her father's sisters); she was forbidden from participating in all activities after college hours, within the college premises or outside, even if they were supervised by a teacher. Her *athai*, apprised of the situation, would be keeping a close watch on her.

Sudaroli was not even in a relationship. Yet the incident had led to the possibility that her higher education would be truncated illustrating the extent to which marriage is considered a compulsory event for young

---

[1] In an article, Kimberly Hart discusses how such personal gifts, which are not meant to be reciprocated, are important in signifying 'modern love', revealing that the male partner is capable of 'investing' in the relationship and taking good care of the female partner by providing for her, even supplying her commodities that were meant to reflect his understanding of the female desire for sensual commodities (2007, 356–357).

[2] Such trends of asking friends to vet prospective partners, before seeking parents' approval is an increasing trend in urban middle-class India, among those of the professional class and the elite, argues Parul Bhandari (2018) in a paper she calls 'Friends are our chosen family'.

women, irrespective of their qualifications or employment prospects. Much of the tension in this incident is linked to the mobile phone—a gadget of anxiety for the freedom of communication it affords young women,[3] which Sudaroli finally used as a peace offering, putting to rest her father's anxieties that she was in a relationship. Sudaroli's 'exchange' of the mobile phone for her 'freedom' symbolizes the choices and negotiations that women must make to enter and stay in higher education, as well as the various contentions around who would be a suitable partner for them, given their family and caste backgrounds, their educational qualifications, their family's aspirations, and how much room they had to negotiate these choices.

'Subjectivities of Suitability'

Given the precariousness of choices and compulsions faced by women in higher education, in this chapter, I examine narratives of intimate heterosexual[4] love in CCT, looking particularly at the role of family, caste, and class in determining the success of choice relationships. Contained in stories of young people's desires, anxieties, passions, and crises, I find

---

[3] Everyday life in the college is significantly shaped by the mobile phone (also see Jeffery and Doran 2013). Today, the mobile has become the equivalent of the greeting card in the 1990s and an older tradition of love letter-writing (see Ahearn 2001; Orsini 2006)—a 'secret space' where certain emotions can be developed that 'cannot be spoken face to face' (Orsini 2006, 239), and indicate a 'willingness to participate in the game of romantic and/or individual advancement, independence and career orientation' (2006, 242; also see Ahearn 2001). Carol Heitmeyer (2016) highlights the role of the mobile phone in contemporary relationships: lovers speak over specially designed schemes offered by mobile service providers. When low on pocket money, they give 'missed calls'. Such a sociality and modes of communicating were also common in CCT. Sometimes, students even changed service providers or bought a 'secret' SIM card to avail themselves of schemes best suited to them. Cognisant of the possibilities introduced by technology and the risk of women exercising their agency to have a 'choice relationship', thus contributing to a sense of loss of patriarchal control, mobile technologies in the hands of young women have become a sore point for many families (and caste associations). In CCT, mobiles phones often arose as a bone of contention between hostel wardens and students: they were seized by teachers and wardens, and young women's 'characters' were often judged by the number of hours spent on the phone. I also heard of families checking young women's mobile phones surreptitiously, checking the content of undeleted messages and 'call histories'.

[4] I did not come across any homosexual relationships in the college. This, of course, does not mean they did not exist, but the insistence on 'homosociality' by the college authorities made the heterosexuality the visible norm by default. It was very common to see young women holding hands, massaging each other's hair, addressing each other by endearments such as *'chellakutty'* (little darling)—all of which was considered acceptable, even desirable, behaviour.

intricate arrangements of relationships, careful strategies and decisions about education and career that determines the 'subjectivities of suitability' (Hebbar 2018) in shaping the course of a heterosexual romance.

The term 'intimate aspirations' suggested by Veena Das (2010) in her essay on a Hindu–Muslim marriage in East Delhi provides a means of looking critically at 'doing love' as one among a set of desires and aspirations. It is intrinsically linked to other aspirations such as status, respectability, good relations with natal and conjugal families, class endogamy, care for self, even as it is a process ridden with anxiety, cris, risk, gendered performance, and careful strategy. Das' invitation to look at love as methodical adjustment and reimagination becomes a useful framework to look at the shifting terrain of desirability that the young women at CCT labelled as 'suitable', reflecting the hope, negotiation, and strategy in the way young people '*do love*' in the college. Their actions and strategies in the face of dire political affect are disruptions of the moral frameworks or rules thrust on them from various quarters, with resistance and conformity becoming part of the process of realizing their desires.

In considering 'intimate aspirations' specifically in line with young women's desire for education, professional qualifications, and class mobility, it is important to locate the plans, actions, and tactics through which young women strategize the success of their relationships, engaging with several aspects of gendered risk. In considering the complexity of these intricate arrangements, I often think of Jaya's case—details of which had been whispered to me over the course of fieldwork as a case of runaway marriage. However, unlike many of her peers, Jaya had not gone 'missing' from regular college life after her marriage. In fact, during the one month that I spent in her class in the Department of Information Technology, I had great difficulty reconciling the image of her presented to me through gossip as the irresponsible young woman who slipped away without a trace causing worry to her parents and loved ones, to the serious young woman who unfailingly sat in the first row, her jaw set in an expression of determination, absorbed in classwork. A year and a half earlier, she had eloped with a man outside her caste (of another middle caste), and twelve years her senior. She simply disappeared one day from the college hostel. No one knew where or why she had gone, not even her best friend, who had been questioned relentlessly by the hostel authorities about Jaya's whereabouts. Parents and police were informed. A day or two later, Jaya returned to her parents' home, saying that she had travelled to a nearby temple town and had been married there with the approval and support of her partner's family. With little choice but to give their blessings to the union, the parents had accepted, and as is the usual protocol during such occasions, hosted a reception for the bridal couple.

News of this elopement had been a huge scandal at the college, but Jaya had appeared unfazed through it all. When she returned to college after her elopement, she had not shied away or become withdrawn because of the gossip that followed her wherever she went. She had held her head high and continued to sit in the first row, engaging in coursework, and retaining her position as the class topper. She had even given an examination the day after her wedding reception in which she had secured the highest marks, and the negative gossip had given way to admiration. Moreover, Jaya conceived soon after her wedding, having a baby over the summer vacation. She then stayed away from college for one semester to breast-feed the baby, coming in only to give her exams. When her classmate Smitha spoke about her, there was tangible awe in her voice: she said that Jaya's highest GPA (grade point average) was in the semester that she had the baby and gone on maternity leave. Apart from the fact that the groom was many years her senior, there could be no objection to such a relationship, many of her friends said while rationalizing her choice to me. 'He is a well settled man (*aal*), with a successful business of his own. Not a college boy (*paiyye*). They have a child also now. She is responsible.'

Reflecting on the gossip around Jaya, it seemed to me that perceptions around her marriage had acquired a sea change: while initially the gossip had focused on the 'respectability lost' and the 'shame caused' by her elopement, subsequently, her acceptance into the conjugal family, the economic viability of the conjugal unit, and her emulation of the patriarchal ideal of a wife who gets pregnant soon after marriage, had recouped her 'respectability' on campus. The age gap between husband and wife also reinforced the legitimacy of the match as it went along the cultural grain of a husband-as-patriarch. It also helped that Jaya retained her position as the class topper—an 'ideal student' when viewed within the pedagogic framework.

Through this analysis, I do not mean to devalue Jaya's resolve to marry her chosen partner; it must have taken great emotional strength for her to strategize her own marriage given the heavy cultural opprobrium against elopement in Tamil society. It must have called for even greater mental and physical resolve to carry and have a baby in her late teenage years, while submitting to the rigours of undergraduate studies and emerging at the top. However, sociologically speaking, what is interesting is the way that feelings towards Jaya's marriage had shifted considerably from shock to admiration among her peers—as she successfully negotiated the requirements of an engineering degree while transitioning to the gendered adult roles of marriage and motherhood. Her respectability, in her peers' eyes, was tied to the performance of gendered roles. Thus, even though Jaya had gone against the norms of a traditional arranged marriage decided by her parents, she had submitted to the gendered 'script' of reproductive labour soon after marriage.

After graduating from the engineering programme, Jaya did not go into employment despite her outstanding grades; she settled into the role of a homemaker after shifting with her husband and child to another city, in which he wanted to expand his business. While one can argue that this was her 'choice', I often wondered what trajectory her life would have taken had she not eloped at a young age. Would she have become a successful engineer? With her sense of stern resolve, could she have been poised at the cutting edge of innovation?

Case studies like Jaya's shows the disruption of a causal line of thinking of the dire effects of a politically explosive marriage—we see the contingent nature of opposition to non-endogamous marriages, rather than fixed ideas of caste and community endogamy,[5] and young women's contingent strategies in such scenarios. It also renders visible the force of the performativity of normative gendered roles, even as young women risk loss of parental support, livelihood, and economic independence in the pursuit of their intimate aspirations. Most of all, it shows how a partner emerges as 'suitable' through the young woman's efforts to ascribe to gendered normative behaviour and self-making.

Jaya succeeded in marrying her chosen partner. Many other women take risks only to be at the receiving end of severe consequences of their family's wrath, forced to discontinue their education, faced with physical assault and violence. Given the contexts of disciplining in the college,

---

[5] This has also been argued strongly by Carol Heitmeyer (2016) in the case of inter-community marriages in a 'divided' Gujarat.

'doing love' was replete with risk, positioning young men and women in deeply vulnerable positions as recent episodes of violence in Tamil Nadu, and the persecution of couples in inter-caste marriages, has shown.[6] While the mainstream media has viewed these incidents as regressive attempts to secure 'caste purity', greater attention must be paid to the historicity of the contemporary moment in which such a consciousness has surged, and the contingent ways in which caste endogamy is enforced (Abraham 2014). In the context of western Tamil Nadu, the political discourse accompanying these episodes has revealed the widespread anxiety around patterns of youth sociality that have emerged in tandem with the upward mobility of Dalit groups, as well as the increased access by young women to higher education and employment.[7] As the targeting of Dalit engineering college students shows,[8] these anxieties are particularly centred around sites of youth sociality such as private engineering colleges that have seen unparalleled growth in the region and appear as gateways of mobility in the social imaginary of different groups.[9]

Perhaps because young women took notice of this, the same interlocutors who urged me to place 'love' as the object of my book also said that they were not interested in contracting a 'love marriage', and sad they trusted their kin to do what is right for them. This allusion to 'love

---

[6] In 2013–2014, the region in which the college is located had been the site of a few well-publicized Dalit murders, termed *aanava kolai* ('insolence murders'), that negatively politicized inter-caste marriage and sociality among college-going students. This had been accompanied by damage to Dalit property and villages. The role of middle-caste organizations such as the Vanniyar Sangam and Kongu Vellala Goundergal Peravai were widely implicated in these episodes.

[7] In caste discourse (such as speeches by caste leaders and pamphlets distributed by caste associations), such misfortunes were characterized as the result of the state "appeasement" of Dalits through affirmative action such as reservation. These killings were further linked with the gendered discourses targeting the social lives of young people in engineering colleges, where young women ostensibly paired off with Dalit men. For instance, in December 2013, at a public rally of 'Non-Dalits', Pattali Makkal Katchi (which has a Vanniyar base) president S. Ramadoss accused Dalit youth of snaring young women from other communities with '… bogus professions of love … they wear jeans, T-shirts, and fancy sunglasses to lure girls from other communities', he said. Citing statistics of a high rate of failure in inter-caste marriages, he had said that it was because 'they were unions born out of caste design and not love'. See B. Kolappan (2012). 'Ramadoss consolidates intermediate caste groups against Dalits'. *The Hindu*, 12 December.

[8] As members of the middle castes and Dalits vie for similar kinds of work in white-collar employment, there is renewed resentment against Dalit claims to middle-class identities through admission in engineering colleges and participation in youth culture. In all three 'aanava kolai' cases that received significant mainstream media attention since 2013, the victims were Dalit male engineering college students.

[9] See Rekha Kaul (1993) and Carol Upadhya (2016a) for similar trends in other South Indian states.

marriage' as being in opposition to 'arranged marriage' carried connotations of love as being transgressive, across boundaries of caste and community, although it was not unheard of to use this term to refer to marriage within one's own *jati*, to signify an agentive decision (cf. Donner 2002). Therefore, letting parents finalize a marriage partner is interpreted as a sign of filial duty, a way of paying back their parents for the 'suffering' they had gone through in order to bring up them well and invest in their education. Such notions of filial duty were reinforced by the imbrication of disciplinary frameworks of home, college, and caste. Thus, while the idea of choice and individualism played an important role in setting up the romantic ideal of a chosen partner, the accounts presented in this chapter reveal that young people recognize that it is kin networks and family support that enables aspiration and mobility. They show great reluctance to completely sever these connections for love marriage, coupledom, and choice. Instead, they speak of strategies to 'convince parents' and get them to accept the relationship rather than state a preference for runaway marriage and risk severing kinship ties.

Sometimes, a relative or one of the parents can even become an ally in strategizing the course of a relationship in order to find the approval of the wider community. One interlocutor, Shanmugapriya had a marriage planned by her mother to a medical college student, which was kept a secret from everyone else including her father. The mother had planned the entire thing when she came to know Shanmugapriya was in a relationship with a young man from the college. She enforced a break up and sealed it with a marriage to the medical college student, a casual acquaintance, who had expressed an interest in Shanmugapriya. She formalized the arrangement in a temple, and told the young couple that there would be a grand ceremony once the groom finished his medical course. Shanmugapriya had no choice in the matter and she went from one surreptitious relationship to another, except that the second had her mother's blessing.

While spoken of as love between parent and child, the discourse of filial piety, especially in the matter of marriage, reveals the difficult circumstances in which upward mobility has been secured for the family and the inherent realization that kin networks are required for success in future careers in the neoliberal regime (De Neve 2016). Such findings also link with studies of marriage among professional engineers in the software industry (Upadhya and Vasavi 2006; Fuller and Narasimhan 2008b; Biao 2008; Baas 2009) and confirm what Kaur and Palriwala refer to as the 'stickiness of rules in intimate practices' (2014, 20).

This is manifested in the way that several young women in the college stated a preference for marriage with close kin or classified kin who would come within the Tamil romantic ideal of *athaipaiyye* (father's sister's son) or cross-cousin, with whom they were close. These kinds of relationships serve a dual purpose: having a protective persona during college life with whom one could experiment romantically and sexually, plan outings, and have intimate conversations. Most times, these young men also had the trust of families, maybe even with hopes on either side that their mutual presence in the college would lead to marriage.

However, it is important to point out that the construction of such an ideal companion within trusted family and acquaintance circles is not experienced by all women in positive and favourable terms. Some saw it as the deployment of family and caste surveillance on their personal liberties and felt that the casting of young men in guardianship roles was akin to having a thorn in the flesh throughout college life. These young men often forced their company and friendship on the young women, even if it was unwelcome. The men often served as conduits to pass information on the conduct of young women in the college to a larger community comprising family, acquaintances, and caste networks, who would then judge the young woman's reputation accordingly.

Apart from these patterns suggesting the 'stickiness' of cross-cousin marriage and caste endogamy (also see Trawick 1990 and Clark-Decés 2014), class mobility also plays an important and decisive role, showing how women viewed intimate desires as being closely entwined with personal, family, and class mobility. Young women tend to see compatibility in educational qualifications and aspirations, keeping in mind the status-raising projects of the family to gauge 'suitability' of possible suitors. Careers in the civil services and management consulting are considered especially prestigious, and usually qualified as a desirable hypergamous marriage. Others insist that the 'shifting circle of endogamy' (Abraham 2014) has widened to include government categorizations as 'same caste' (between different castes classified as OBC, for instance). Many said that it would be difficult to 'convince parents' on the suitability of matches that are outside these broad brackets.

Apart from loss of face in the community and the threat of caste-related violence, it would also affect the economic balance in marriage, the young women said. For instance, a non-desirable marriage may result in non-payment of dowry, which many young women considered important for their self-respect in the marital home. Such preferences shed light on the 'economies of love' (De Neve 2016) in South India: not just in terms of the necessity for kin support in the neoliberal regime, but also practices connected to what M. N. Srinivas (1984,13) refers to as 'modern form of dowry'. This form had intrinsic links to the 'professionalization' of communities that once paid a 'bride price'. According to Srinivas, in the late nineteenth and early twentieth centuries, several non-Brahmin castes started paying a 'dowry' to 'desirable' grooms who had broken out of village agrarian life and found a foothold in the urbane formal economy either through a profession such as engineering or medicine, or a job in bureaucracy. This 'modern' practice reflects the value attributed to a job in the formal economy as well as pride in the growing stature through the making of powerful connections, especially given the pressures of *jati* endogamy (Srinivas 1984; also see Biao 2008).

Therefore, what may be perceived as a 'paradox'—educated young women expressing the need for dowry—reflects what the sociologist M. N. Srinivas (1984) had noted women having internalized the need for dowry, casting their own subjective notions of self-respect to the value of commodities that they would be able to take to their conjugal home. The degree of internalization is so intense that the need for dowry appears *sui generis*, even when *jati* endogamy is not at stake. It also reflects the continued value placed on 'engineering degrees' in middle-class contexts, and the expectation that a daughter with a professional degree should at least be able to snag a groom who has a professional degree himself. Also, this enables parents to 'handle' the relationship as if it were arranged, without loss of face, as Henrike Donner (2002) finds in the case of what she calls 'self-arranged marriages' in a middle-class neighbourhood in Kolkata.

In deeming whether a match was suitable, some were concerned with the status of their partners, while other young women discussed how achieving a certain degree of status themselves could lend acceptability to their choices. A student of Fashion Technology, Arasi accepted Raj's proposal in 2014, three years after he first declared his love for her. Back when they were in class twelve, Arasi had rejected his proposal saying she was too young and was afraid of what her parents would say. However, Raj had persisted with his overtures, including giving 'missed calls' to her mobile phone to let her know he was thinking of her. This had come to her parents' notice, and they suspected her of reciprocating his feelings, and being in a relationship with him. They had beaten her for getting involved with a young man instead of focusing on her studies. Depressed and angry at this turn of events, Arasi had slit her wrists, the scars of which were still visible on her forearms—physical reminders of the harm a problematic relationship might bring.

Yet Arasi finally accepted Raj's proposal when the two lived in separate cities, studying in different colleges, saying that she had begun to reciprocate, and had also worked out a strategy to 'convince' her parents. She said:

> I want to pass Civil Service (examinations) and become an IAS officer for the status. Having a daughter who is IAS is a matter of *geththu* (status) for the parents. They have to accept our love after that. I will take training … IAS coaching from Delhi, so I can pass the exam. I will not sit for any of the other exams—only for Class I grade exams—I will try my best. I have already started classes …

Arasi's plan highlights the importance of status and class in parental assessment of whether a marriage is considered suitable (or not). Convinced of her plan, when I visited her room in the evenings after class, I would find her reading the newspaper, making notes, and looking up various coaching centres in places such as Delhi and Chennai to prepare for the civil service examinations. Interestingly, when I asked her a few months later about her preparation for the exams, she told me she had decided to put it off for a few years as her caste group (the Kamma Naidu/a General Category group) had begun lobbying for Backward Caste status in Tamil Nadu. 'There has just been a state-wide meeting of Kamma Naidu in which it has been decided that we will fight for BC status, and I am going to wait … my chance of cracking the exams will be higher if we secure BC status', she said. Her revised strategy for marriage involved spending a few more years in higher education so her partner could 'be settled' by the time her parents raised the question of marriage and began looking around for a suitable groom. She enrolled in a MTech programme and then a PhD at a university in Chennai.

However, caste and caste networks continued to pose a problem for their relationship in ways the young lovers could not have imagined. Arasi's relative who had been classmates with the couple in school had come to know of their relationship from friends, and informed her family, and Arasi had been categorically told by her family to break it off. Arasi tried convincing her parents about the suitability of the young man only to receive jolting news from him that his elder brother had been implicated in two cases of caste-related murder in their home town in Tirunelveli district. Her parents felt vindicated by their rejection of Raj, and stressed the need not only for caste endogamy but also regional endogamy, highlighting that any boy from the 'violent south' would be incongruous with their ways and manners. It would be better to find a Naidu boy from the Kongunadu area, they had said. Given the seriousness of the crimes in which Raj's family was now embroiled, Arasi too found herself 'choiceless' in the matter of asserting her right to continue the relationship with Raj. She felt that eventually her parents would find a groom who would suit their criteria and she would have to be married to him. She continued to resist, however, by asking her family to let her focus on her studies instead of pursuing alliances for her marriage. This indicates the many ways in which education has emerged as a tool for many young women to 'buy time', until a suitable solution can be found to their dilemmas.

Gendered Risk, Sociality, and Modes of Courtship

While marriage strategies, the enhancement of status, and the reconciliation of parents' desires to one's own emerged as a central component part of the pursuit of 'intimate aspirations' in college, one cannot reduce young people's 'intimate aspirations' to the marriage framework alone. Everyday life in the women's hostel is suffused with the language and performance of heterosexual desire. Young women tease each other about boys, discuss the intricacies of sex, and speak about their intimate desires. They are also curious about the objects of each other's desires and *sighting*[10] young men

---

[10] 'Sighting' is essentially an act of ogling someone, not lecherously, but in a manner that is often welcome and considered flattering. It could mean checking out someone, or a long-sustained infatuation or flirtation.

is an essential part of everyday life. Though students often dismiss these acts as 'timepass' and not a 'committed relationship' (also see Abraham 2002), such relationships can also involve hours of texting on the mobile phone and arranging to meet 'by chance' at various locations on campus. Young couples do not 'date' as the term is popularly understood. They refer to themselves as 'committed' to suggest that they would ideally like to get married. When they meet, it is usually to take a ride together on a bus or in semi-secluded venues such as parks in remote areas of the city. Those who could afford it went to 'fast-food joints' that had cleverly designed architectural features that could give couples some privacy—semi-lit basements or dark mezzanine floors with booth seating, separate from open-plan seating for families. These activities thus show the way 'intimate aspirations' emerge in contingent ways, broadly falling into what young people may categorize as wanting to 'enjoy' college life, carrying connotations of engaging in pleasure, experimenting, experiencing, cultivating a relationship, and remaking the self in light of such experiences. In other words, heterosexual attraction appears as an important event in women's lives in a multiplicity of ways—it is an event that marks their biographies in definitive terms, inscribing them with new knowledge, new feelings, and new ways of thinking about themselves.[11] 'Love' alters states of mind, habit, and brings about attachment. It is characterized by thrill and excitement as much as pain and disappointment. Young women visibly crave the company of their partners, attempting to cultivate a companionate intimacy, earnestly waiting for them to text or call, and shedding tears when they argued or quarrelled. Moreover, to the question on what I, as an ethnographer, should study about them, I was always told 'love life' or 'relationships'; it seemed to them that it was the only aspect of their lives that could be interesting for anyone.[12]

---

[11] Alan Badiou with Nicholas Turong (2012) illustrates the process of love as a construction that involves remaking the self as two, not one, in the philosophical mode, in *In Praise of Love*. Badiou's work has been invaluable in analysing 'love' as a process of construction and love as being an event.

[12] I also meant this question as an ice-breaker and my interlocutors' responses were also meant to test the waters. The fact that I looked interested rather than intolerant of such responses cast me as an empathetic figure. I am sure that my own age at that time (26–27) and my status as an unmarried female student who was not from Tamil Nadu, helped my case. I was widely perceived as an 'Akka' (older sister), who was experienced in such matters (I was widely questioned about my own romantic and conjugal involvement), and would not judge them harshly for their choices. The young men did not trust me as easily, and for a while, rumours circulated that I was the

The crux of the intimate aspirations of many young women, therefore, may not be to carry out a serious relationship or find a marriage partner, but to look good, or present themselves in ways that people find desirable, and engage with risk in the process. Shobhana, a student of Fashion Technology at the college, was a resident of Salem City. Educated at an elite convent school, she could articulate well in English and came to college dressed in the latest cuts and styles, wearing startling green contact lenses, framed by heavily lined eyes. In my conversations with her, it emerged that she saw her fashion sense as being tied to her aspiration to become a fashion designer. Her father, a professor at the university nearby, had explicitly forbidden her from taking up fashion design, and told her she had to join an engineering college. She agreed, but insisted on enrolling for the Fashion Technology course at CCT—the closest cousin to fashion design within the engineering bouquet. She was disillusioned that the course did not have design as a central component, but attempted to keep in touch with fashion trends around the world. Reflecting her interest, she built up a collection of accessories to complement what she called the 'dull' and 'boring' clothes prescribed by the college dress code; her accessories included dramatic bows for her hair and trendy jewellery. Her signature, however, was the green contact lenses, which had become something of a sensation in the college: others often discussed it; some even imitated it. It reaffirmed for Shobhana her ability to set trends and emerge as a 'fashion icon'.

Her attention to her clothes and appearance, however, had earned her the reputation of being a '*scene*'. A derogatory term signifying someone who hankered for visibility and attention, it also carried connotations of class—having a taste for the latest fashions, high investment in clothing, usually afforded only by the elite, and seen as a sign of 'loose morals' advertising sexual availability.[13] Shobhana was aware that she was targeted often for comments and slander. Her tag of being a '*scene*' had many repercussions, she said.

> People say I am putting a 'scene'. I have to face a lot of problems because of this. I cannot bear the number of 'Friend Requests' on Facebook, and

'principal's spy', which I quelled by asserting my identity as a student, with the help of the young women interlocutors who had accepted me more easily and met me more often at the hostel.

[13] Also see Nakassis (2010, 128–136).

I cannot put up pictures of myself because everyone downloads these pictures on to their phones and saves them. So, I changed my name [on Facebook] to Sammy Barbosa. The next day, when I walked past the [male-dominated] Mechanical class, I heard the whole class shout 'Sammy Barbosa!' and heckle and laugh. This happened less than 12 hours of me changing my name. I went back home and typed my name—Shobhana Balaguru—and there were two other fake accounts in my name, with all the correct details. The fake accounts had about 450 friends. I don't know; that person must be chatting with so many people under name, but what can I do? Some boys follow me home sometimes as well. But what to do? I cannot even discuss these problems at home. If I do, they will probably pull me out of college and not let me go out.

One way of reading Shobhana's experiences is as posited by Martyn Rogers (2008), who observes that subaltern youth in the city of Chennai legitimize harassment, especially of articulate English-speaking, convent-educated young women as being 'deserved' because of their 'modernity' and resultant 'loss of modesty'.[14] On the other hand, Osella and Osella (1998) suggest that 'harassment is only half the story', and 'idealised romance is equally important' (1998, 192), and compel us to think about the links between harassment, flirting, aggression, and love, and the difficulties in differentiating between them. On this point, I agree with the Osellas—despite the mutual hostility between the sexes (as evident in the previous chapter), all youth are keen to initiate relationships with the opposite sex, to text and chat, to flirt, and begin romantic relationships, even if for *timepass*. In the sex-segregated milieu of the college, such aspirations are hard to realize, and what can be construed as 'stalking' or 'harassment' often becomes a legitimate mode of expressing interest, which is also acknowledged and interpreted as such. Thus, even something such as 'eve-teasing' can also be located on a system of graded hierarchies—some forms of which women enjoy and participate in, while more violent forms are experienced as 'harassment' (also see Krishnan 2015, 127–133). This is not to devalue Shobhana's experience as a 'victim' of 'harassment' or her feelings of vulnerability or abject helplessness, when a bunch of

---

[14] Rogers (2008) views sexual harassment as a somewhat unintended, trickle-down effect of the competition for white-collar employment.

men followed her home, or trolled her online, especially when no mechanisms of support or redress exist. However, rather than favour a mode of analysis that constructs Shobhana as a passive victim, it is important to acknowledge her *style* in ways that accurately posits her agency. Sneha Krishnan (2018, 56) describes such fashion decisions, among 'bold' college-going young women in Chennai, as a way of 'doing youth' and of gaining attention and becoming popular among peers—not just to start a romance but also to question flirtatiously young men's exclusive and often gendered occupation of public space and youth culture.

However, being a 'bold' young woman might have its own consequences. Monika often found herself vacillating between embodying respectability and enjoying the 'playful' phase that is supposed to characterize the period of youth and college life. In her first year at CCT, she identified more closely with the 'NRI' gang[15] (as she was from Hyderabad, the cosmopolitan capital of the state of Telangana) and hung out with them. She enjoyed outings with them, sometimes to very elite spaces such as a sports club, walked the runway for a fashion show featuring 'Brides of India' during a college event, and tried to 'enjoy' college life to the fullest. However, she reflected that such a lifestyle did not bode well for her in the long run, as she found 'local' students cultivating a distance from her, wardens profiling her, and men sending lewd comments in her direction. In the second year, she decided to tone down her lifestyle, learned Tamil, hung out less with the 'NRI' gang, and more with her room-mates and seniors, whom she addressed with the Hindi '*Di*'—a fashionable spin on the Tamil kinship term '*akka*' generally used to refer to seniors. She used her sense of fashion (and skills with the dryer) to style friends' hair, used her English skills to translate foreign films for them, and entered the economy of affective kinship with Tamil seniors and classmates. When asked why she never participated in the fashion show in subsequent years, she said, 'Never again ... I cannot handle it', referring to the surfeit of male attention that followed. However, she still entertained her friends by regaling them with stories of her many admirers in college (whom she called her 'fans') and their light exchange of banter. In fact, many a

---

[15] NRI stands for 'Non-Resident Indian'. However, this term is not be taken at face value. In common parlance, it refers to anyone who is not from Tamil Nadu. In CCT, the term 'NRI' was also used to describe students from other parts of South Asia and Africa. It carries connotations of being 'elite' and 'high class', of someone enjoying a 'globalized' lifestyle.

time when I walked with her around campus, we would hear the Telugu words *'Jargandi! Jargandi!'*—'Make way!' 'Make way!'—muttered by the young men standing on the sides of the road or the corridors, alluding to her Andhra roots, and her last name being the same as the presiding deity in a popular temple where these words were used in order to rush pilgrims to complete their *darshan*[16] in a few seconds and move away. The performance was meant to show they admired her—even a veiled positioning of her as goddess giving *darshan*, and they as devotees. Such playful episodes flattered Monika, and she loved describing them to her room-mates and friends once she was back in her hostel room.

However, Monika never told her friends when she started developing feelings for a young man from a neighbouring state. She said she was too scared that she would share the same fate as one of their other friends, Sasikala, who was ostracized from their group after they found out that she was 'too close' to her boyfriend. In this way, Monika worked herself out of the 'NRI' label, and its 'high class' and 'cosmopolitan' connotations, to cultivate the image of being a 'good girl' (*adakamaana ponnu*), of being modest. So though 'bold' and '*stylish*' women often use their agency to negotiate friendship hierarchies and open liminal zones of enjoying a youthful lifestyle, usually open only to men, it must be located as a fluid identity, often twinned with 'a good girl' label that is equally driven by agency. In fact, constant labour goes into reiterating mainstream discourses and performing modes of gender in order to find acceptance within hostel cliques, to stay in engineering college, and in time, enter paid employment. Like the college authorities, young women were acutely aware that their youth had to be 'managed', so they could stay in education and enjoy the little pleasures afforded by being away from home.

Thus, despite the rules aimed at segregation and the constant surveillance from various quarters, many young men and women engaged with each other playfully, passing comments and retorts as they passed each other, or met each other on campus. To an outsider, such interactions may not even be visible—as they may be contained in a quick wink, a

---

[16] '*Darshan*' is the process of seeing and being seen by the divine—in which a viewing event is meant to establish 'divine contact'. Diane L. Eck (1981) lists the many qualities of *darshan*, as relating to knowledge, philosophy, and insight in an eponymous volume. Christopher Pinney (1997) explores the term further to examine local ways of seeing.

hand gesture, a fleeting touch, or a simple nod of the head, as people passed each other by. I remember being baffled when I first started accompanying young women on their 'meetings' with young men. I fully expected that we would stand around and chat with the young man concerned, except that we would just pass by each other and the 'meeting' would be over even before I realized that it had happened.

Such interactions built up a particular charge in spaces policed by security guards who ensured that young men and women could not sit down together for a meal. In my experience, young women appeared to ignore these comments while strategically deploying silence, or giggling and acting hyper-feminine, or occasionally hitting back with comments (*comment adichu*). These modes of sociality also marked the beginning of romantic liaisons for many: following a chance conversation at a bus stop, a young man and woman might become 'friends' on Facebook, start chatting on private messaging services on social media or on the phone, propose, before claiming to 'love' each other. Such 'texting' relationships exist not just between students, but also between students and younger members of the faculty and the external 'trainers' who came to coach students on how to train for successful recruitment.

What such incidents also show is young women's everyday negotiations between risk and pleasure in the pursuit of their aspirations for self—whether they want to dress in a particular way or enjoy playful banter with members of the opposite sex. 'Flirting', 'friendship', 'love', 'timepass' emerged as ways of negotiating the risk–pleasure binary that forms the central scaffolding of their lives. The meanings of such relationships were constantly negotiated, renegotiated, remade, and redefined. For many, such engagement was also passive acquiescence that they might never be able to negotiate a choice marriage, but to enjoy the time in higher education engaging in short-term relationships. This did not mean that such engagements were, in any way, superficial or bereft of meaning: as underlined in the preliminary notes of this chapter, such attractions and intimacies emerged as very important events in the lives of my interlocutors. It directed their gendered performativity, subjectivities, even their self-making in the 'shifting exigencies' of the 'everyday' (Kleinman and Fitz-Henry 2007). In unravelling the many meanings and processes through which these young women made sense of their intimate relationships, I try to understand what it means to be in love, be attached, to be involved, to suffer a heartbreak, and to move on. To engage in such aspirations was to also engage in adventure, transition, change, affect, and to open oneself to vulnerabilities and failures.

Love, Self, and Shifting Meanings

A student of the IT stream, Reshma was referred to as a '*figure*' on campus because she was widely considered attractive by both men and women. Unlike '*scene*', which is a derogatory term with hyper-sexual, being a '*figure*' is an aspiration in femininity and heterosexuality; it referred to finding a balance between being attractive, but also demure and respectable. With her long straight hair in a 'laser' cut (cut in layers), pastel salwar kameezes, her affable ways, and good grades, Reshma came close to embodying this ideal, and had several admirers, some had even proposed to her. When I asked her if she took any of these 'proposals' seriously, she said that such things were fine and fun, but she had a strict 'three-month policy' during which she would let herself get close to any of them—after which she would consciously distance herself lest her feelings become 'too serious'. She was not interested in 'commitment'. However, I witnessed how Reshma could not stick to her own 'policy' when she got close to Deepak, a student from a South Asian country, and how three months turned to six, and then nine. Throughout this period, she text messaged Deepak everyday, developing a close intimacy with him; she updated him on the day's events, felt hurt when he did not respond to her text messages, and spoke a lot about him to her friends and me. With him, she discussed possible adventure—the possibility of trying out alcohol for the first time, the possibility of a getaway, to visit him in a neighbouring town where he was admitted for surgery, and maybe even visiting his country.

He was very 'possessive' and demanding that she discuss every aspect of her life with him, and make decisions keeping his opinions in mind. Thus, once when she went to a mock aptitude test at a coaching centre in the city without telling him, she found a text message asking her, 'Why didn't you tell me that you were going to answer this exam?'. One of his friends, who was also attending the test, on spotting her in the exam hall, had sent him a message saying 'his *figure*' was taking the exam too. Despite telling him many times that it was a rushed decision taken only that morning, he continued to be upset and angry with her. Even though theirs was a 'timepass' relationship, he demanded accountability on the basis that theirs was an intimate relationship in which 'they told each other everything!' (cf. Giddens 1992). She spent the rest of her day anxiously trying to mollify him.

Witnessing such incidents, her friends asked if this relationship was 'serious', but she always responded saying it was nothing but 'timepass'. However, one evening Reshma returned from her class furious with her room-mate, Haripriya. She told her other room-mates:

> What does she think of herself! Why is she texting Deepak? She got his number from me, and now she is suddenly sending pictures of herself to him. I have to ask her, 'Do you love him?' Why hasn't she told anything about him to us? She knows I text him, and then sending a picture like that to him. Only 3–4 days she has texted him. And then sending this picture. What does she think of herself? She has become a joke. Sending a picture like that to a boy she barely knows. In these 2–3 days, has she fallen in love? Does she understand what these boys are like!

Once she calmed down a little, the details became a little clearer. Her room-mate Haripriya had started a texting relationship with Deepak after Reshma had borrowed her phone to text him and accidentally left his number saved on the phone. Just three or four days after they started talking to each other, Haripriya had shared a picture of herself with him; in the picture, Haripriya was wearing a new skirt with a short top and a stole wrapped around her shoulders. Such clothes were prohibited by the dress code and were only worn within the precincts of the hostel, which means young men never saw them in these *avatars*. Sharing such a picture with a young man was considered an invitation to intimacy.

Deepak, in turn, had shared this picture with Reshma, who got furious with Haripriya. When Haripriya came into the room a little later, Reshma confronted her. Haripriya defended herself saying that she thought Reshma was just doing 'timepass' with Deepak and had never claimed any special status for their relationship. 'Don't you love Nishant?' Haripriya asked, referring to a young man from Reshma's high school. I had become familiar with this name because Reshma often spoke of him too, of the many glances they shared during their school days, the phone calls, the meetings, even the music that had played when they met … her descriptions had carried a cinematic quality, and we had all listened enraptured. Haripriya, too had believed that Reshma still carried a torch for Nishant, and had not been aware of crossing any boundaries when initiating a 'texting relationship' with Deepak, she said. As far as she understood, Reshma loved Nishant. However, as Reshma's outburst reveals, she had become very jealous. After she narrated her story to her other room-mates, they also agreed that Haripriya should have known her boundaries and stayed away. Neither Reshma nor her room-mates pointed out that perhaps Deepak, too had a role to play in encouraging the intimacy Haripriya had sought. After a protracted quarrel between the two women, Haripriya finally gave up texting Deepak and an uneasy relationship resumed between the two room-mates. Reshma and Deepak continued their texting relationship; the jealousy that had flared in Reshma forced her to be more honest to herself about the way she felt about Deepak. Over the next few months, she continued the charade of pining for her high school sweetheart in front of her room-mates but grew much more intimate with Deepak. She met him sometimes, without telling them, visiting places such as the park or the zoo. When I asked why she did not tell her friends, she said that they would get judgemental, but she could tell me because I was like an *Akka* [elder sister], who understood relationships [were complex], and could be trusted to not gossip with her friends and raise a scandal.

Moreover, an event occurred a few months later in the same college that made her introspective about the prospect of taking the relationship forward. This event was the wedding of a Nepali male student and a young woman who belonged to Salem city; both students were classmates in the MBA course who had met at the college and fallen in love. The campus was agog at the impending marriage; everyone seemed to be talking of the wedding and the acceptance from both sets of parents, from very different communities, that the marriage implied.

The wedding was to be an early morning affair in a nearby temple town, with a grand reception to follow in Salem, hosted by the bride's parents. All the (foreign) South Asian students in campus, including Deepak were invited. Reshma received a picture of the bridal couple soon after the *muhurtham* and forwarded it to me—the bridegroom was dressed in a white veshti with a slim gold border, white shirt, *angavastram* across his shoulder, and wearing sunglasses. Befitting the simplicity of a *muhurtham* in a temple, the bride wore a pale pink cotton-silk sari, red blouse, and was adorned with gold jewellery. Reshma sighed at the happy picture, especially at the sight of the Nepali bridegroom in a South Indian costume. She told me that after seeing the photograph, something had changed inside her, and that she had been quarrelling endlessly with Deepak over the past few days—'why can't we do it, too?', she had been asking, pestering Deepak for details of the bridal couple, how they told their parents, what they said to convince them, and so on. Finally, Deepak told her that the groom's parents in Nepal did not even know that their son had married! He had gone ahead with the marriage without even informing his parents, although the young woman's parents had been aware, consented to the relationship, and performed the marriage. The bridal couple would even be living with them after marriage.

I had no way of ascertaining if this was true or if it was Deepak's way of putting an end to Reshma's hopes that his parents could be convinced about the suitability of South Indian brides, but it certainly dashed Reshma's hopes. As she sat and discussed the feasibility of the match from the point of view of Deepak's family, I quizzed Reshma about how she would get approval for the relationship at her own home, suppose Deepak were to agree: 'they will kill me', she said, mimicking a blade slitting her throat. 'Not my father, but certainly my uncles. I am so scared to even wear sleeveless clothes or a short top in front of them'. Yet she was trying to plead her case, 'He is brilliant, a topper in his country, the only one who cracked placements in X company. I don't know if I will get another boy like him', she rued. 'He is good looking too, maybe my parents can accept him like this girl's parents here [referring to the newlywed couple]'.

In Reshma's 'three-month' policy, we see young women's contingent strategies in inaugurating zones of romantic experimentation and sexual initiation for themselves. Despite setting herself up for a 'timepass' relationship given the sanction against love marriage in her home, Reshma found herself slipping into a zone of deep affect, nursing marriage aspirations even in the face of possible assault from her family. Despite such thoughts and strong feelings, Reshma continued to perform the relationship as 'timepass', cracking dismissive jokes about Deepak in front of her friends. When she began to meet him privately, she shared it only with one or two friends, rather than her entire peer group for fear of censure and sanction. Such states of liminality and desire at crossroads are important factors when considering youth relationships, rather than a typology of relationships, or classifications of sexual and platonic (cf. Abraham 2002; Clark-Dećes 2014,157; Osella and Osella 1998, 196). Desire changes over time, it is subject to changes in the immediate environment, the powerful influence of images/imaginaries, and peer pressure as much as family and kin approval. Agential decisions are weighed: good looks, professional competence, and notions of companionate models are weighed against notions of caste and kin approval.

When Deepak and Reshma finally broke up a year later, it was contingent on multiple factors: one, Deepak, being Reshma's senior was about to graduate. Two, on one of her visits home, Reshma's older sister had read the text messages she had exchanged with Deepak, and had informed their mother about Reshma's possible romantic involvement. Livid, her mother and sister had confronted and reprimanded her, and then completely stopped speaking to her, making her life 'intolerable'. 'They don't talk to me properly, and when they do speak, it is only to taunt me. I just cannot bear it', she confided to me over the phone one day, her voice choking:

> Deepak too cannot take any decision, and he says once he goes back to Nepal, he will not be able to talk often on the phone as it is too expensive, and that we won't be able to talk ... When I ask him how he can just leave and go, he says 'what has happened between us? Nothing has happened between us'. All these months, all these months ... he just used me for timepass.

As I struggled to console her, I found myself searching for traces of the young woman who boisterously once told her friends she was only doing 'timepass' with Deepak. She had wanted no residue of intimacy from any college romance, her 'three-month policy' was meant to assure that, but found herself nursing a deep gash—deeply pained by both Deepak's juridical denial of their involvement, and the sense of rejection she was facing at home. 'I really miss him, man[17]', she said, sobbing.

---

[17] 'Re' is a casual reference to the person spoken to—like "I really miss him, man".

Understanding Reshma's 'intimate aspirations' of following a 'three-month policy' requires reflection on the performance of gendered subjectivities in college. Reshma, it seems, was able to briefly invert patriarchal norms of passive femininity, attempting to live out a fantasy of multiple romances—'No Strings Attached', to quip from a Hollywood movie she often watched, while acting as translator and passing on worldly knowledge to her more 'innocent' friends about what it meant. It was essentially a masculine fantasy within the Tamil context—of having fun, enjoying, while making no 'commitments' (see Nakassis 2016; Nisbett 2009). In the college context, it emerged as a stereotype often attributed to the foreign students on campus, the people who ostensibly did not understand the value placed on chastity. However, over time, such a fantasy seemed to have been replaced by one of romantic love leading to marriage and migrating to a country she had never visited, and whose culture she did not even understand. She made sense of such a relationship by looking at everything they had in common—engineering stream, academic background, good prospects in employment—compounded by the physical attraction she felt towards him, the sense of intimacy inaugurated by telling each other everything over a period of many months, and the feeling of shared adventure in facing risk. In recalling the many months that she had spent with him, she was not referring to physical intimacy, but the companionship he had afforded by his many text messages throughout the day: words to which she had become habituated, reading them last thing before going to bed, first thing upon waking up, and at regular intervals through the day, her mobile phone clasped in her hands even during class. In missing him, she alluded to the sense of self that he had helped her feel; that sense of self which had taken her away from her stated 'three-month policy' of having an attachment to simply pass her time, to a place of meaningful involvement and deep affect.

Reshma's short-lived inversion of prescribed gender roles, and later hopes of a romantic level were thwarted by Deepak's final denial of involvement which reduced her, in her own head, to the stereotype of the 'local' young woman who got carried away by a foreigner's charms. Updates from her other friends dampened her spirits further: she learned that Deepak married his childhood sweetheart as soon as he went home after graduating. 'I just can't believe he used me like that', she said during subsequent conversations. 'He never even had the intention …', she said, sounding disappointed and hurt.

Almost a year later, I received another call from Reshma:

> I am going mad … Nishant said 'no' … (I am) feeling so hurt. I was always okay with what happened with Deepak because—heart of hearts—I always knew that Nishant is the one I am going to end up with. I always thought Nishant is the one for me . . . because we always had that connect. Such an awesome guy, man. Even if we did not keep in touch, even if he did not call … I knew somewhere that he is the one. That phone call once a year on my birthday was enough! That was enough to tell me that one day we would be together, and that we need not be in touch everyday in order to have a future together. But today, I was talking to my old friend; she mentioned he is going to Japan … he's got some prestigious scholarship. You know he studied in IISc (Indian Institute of Science—a premier research institution in Bangalore), no? I called him, I asked him when he is leaving for Japan, for how many years, and then I casually asked him 'what are our plans?' He said 'what plans?'. He never had any plans of being with me. He said that he always knew I liked him, but he is not what he used to be in school, and I don't even know him and I cannot possibly love him. He tried to make me understand. I told you he was a nice guy, no? I don't know what to do … I got such a shock! You know I always loved him, no, man!

I narrate this incident not to falsify or render untrue one narrative or another, or even point out certain contradictions, but reinforce the contingency and construction of love and romance in the narrative, and the subjective construction of self. In Reshma's narrative, her love for Nishant was posited as a 'true love' as opposed to the shifting meanings attributed to the 'timepass' relationship she shared with Deepak. In talking about Nishant, she was able to summon a sense of self that was deeply attached to him, with specific meanings attached to his presence at various events in her life (like his phone call every birthday). In talking about her connection with him, she seemed to resuscitate what she believed was an authentic self—one that had been pristine even through the time that she had felt deeply for Deepak. The making of the self then emerged as a subjective response to whatever is at stake at a particular moment—what Kleinman and Fitz-Henry (2007) calls the malleability of 'lived experience'—conditions that remake memories, emotions, cognitive style, and even the deepest sense of self. As the case study shows, one cannot subscribe to atomistic ideas of 'love' and 'individual', but focus on contingent self-making given the shifting conditions of human life.

This is highlighted further in the turn of events in Reshma's family a year later when Reshma called me with a request: 'Can you please help me write a resignation letter?' she asked. It was not out of the ordinary for me to receive such calls from my CCT interlocutors who regularly sought my help with writing official letters or drafting statements of purpose when applying for Masters programmes. I considered it the least I could do in return for how they had shared their innermost thoughts, hopes, aspirations, and desires with me.

'Sure', I said, thinking that Reshma wanted to quit her own job, about which she has been voicing her dissatisfaction to me over many months. 'Congratulations on finally deciding that you will quit! I am sure there are other, brighter opportunities for you', I said. She replied:

> No, re. This is for my older sister. We have just found out she is in love with her colleague—a Malayalee—and my parents have forced her to come back home. My sister is fighting saying she wants to marry him only and she is also upset that she is being made to quit her job without serving the notice period, and I have told her I will try my best to at least secure a relieving letter from her company so all her years of work experience do not go waste'.

'Oh my god! Is there no way you can convince your parents to let her continue working?', I asked.

> No, re. It is better she comes home! She is really immature and I know this colleague is not a good match for her. I know my sister inside and out, and I can tell you that their marriage will not be successful. He is not man enough for her. I have always told you that my sister is timid and shy, she needs someone who will fend for her. I have met him—there is nothing in him! If she continues in the same office, what sort of a break up will it be?

'Er ... but maybe she knows him better ... it must be so terrible for her to be brought home like this, and forced to quit her job. She has an engineering degree too, just like you', I stammered, struggling to understand what position I must articulate. Should I stand true to my political beliefs or should I blindly side with Reshma in order to remain her confidante?

'She does not know what is good for her', Reshma said, sounding diffident now, and I thought back to the time when the same Reshma had helped her sister 'run away' from home to avoid a family who wanted to come and 'see her' (*ponnu pakunu*) for a marriage alliance for their son. The sister had come to CCT and bunked in Reshma's hostel room for a few days, until their parents had made some excuse to the prospective groom's family. This incident had been repeated to me many times over as an instance in which Reshma had to be vocal and ambitious on her sister's behalf, as her sister did not always know how to fend for her interests. However, Reshma and her sister now seemed to have switched positions and Reshma was penning a resignation letter as her sister was kept under 'house arrest' in a way that was reminiscent of her sister's behaviour the previous summer when she had come across Reshma's messages to her then boyfriend, and informed their parents, landing Reshma in trouble.

Reshma's case shows not just the importance of suitability in deciding the match, but the ways in which the patriarchal structure of the family exerts itself even through its youngest members, like siblings and cousins who might belong to the same age cohort, to contextualize and decide 'suitability' incorporating them into acts of policing and disciplining young women and undermining their choices.

However, family dynamics are also continuously made and remade, and Reshma's sister's 'fall from grace' spurred the family's aspirations for Reshma. Knowing that enforcing a break up and convincing the sister to marry someone else may take time, delaying not only her sister's marriage but also Reshma's, they encouraged Reshma to start looking for opportunities in the United States of America. Within the culture of migration in Andhra Pradesh (Upadhya 2016a), this seemed like a respectable path for buying time—a decision that could mask the contentions and complexities within. Within a couple of months, Reshma was accepted by a little-known private university in New Jersey and flew the nest, even as the family took on greater financial stress to enable her move.

Gendered Dynamics and Relationship Trouble

However, it is not only the family that represents risk and exerts its disciplinary apparatus over young people's autonomy and choices. Young men and women also make and remake each other in contexts of love, romance, and transition to adulthood. While some may find choice relationships liberating, models of romance are also heavily influenced by cultural and gendered stereotypes of inside–outside spatiality, dress conventions, and possessive behaviour.

I first met Rathna when I was visiting some other interlocutors in their room. Her friends took pride in introducing that she was a former sports secretary, and a member of the college throwball team, often called to represent her zone and district in various prestigious tournaments. Her room had been full of medals and trophies at the end of the previous year because of the number of tournaments she had won, they said, as Rathna laughed and brushed aside her friends' praise. Though Rathna had seemed embarrassed initially with the teasing, over the next couple of days Rathna hung out with me and shared details about her life—she had gained admission through the sports quota, and most of her life had been spent on sports grounds. Her mother was a school teacher and her father ran his own business. She was also in 'committed' relationship with Kamalesh, a young man from the same stream as her but in a different section—she showed me a picture of him on her phone. She hoped to marry him one day, but expected that her parents would oppose the match. 'We are from different castes, and that is sure to upset my parents', she said. 'But I still hope to convince them'. She was excited to chat about Kamalesh, sharing details of their growing intimacy such as when he made her meet his sister, bought fabric for a salwar kameez for her, or changed his profile picture on social media to a picture of only her eyes (not her whole face, to preserve anonymity) as a dedication on her birthday. I also found her cooking chicken one day, explicitly against the hostel rules. Because she craved chicken, Kamalesh had brought some meat for her, with all the ingredients required —tomato, onion, ginger-garlic paste, some powdered masala—to make a chicken curry.

As months passed, however, it appeared that sports had begun to take a backseat in Rathna's life. On separate occasions, she discussed different reasons for her withdrawal. At first, she blamed general academic anxiety—with all the classes she had to miss because of tournaments and practice sessions, she was worried of not doing well in examinations and not preparing well enough for campus placement. She also asserted that part of it was parental pressure: she discussed how her parents had never been supportive of her playing sports in college, though they had encouraged her a lot in school. They were concerned that excessive time in the sun or injuries can mar her looks and take its toll on her health and physique, in turn affecting her marriage prospects. She also began to discuss the difficulties of women in sports: she said that the young men often stood around leering, commenting on a young woman's physical features and playing style. 'It often gets very uncomfortable out there, even though we give back properly to the boys!'.

A few days later, as we hung out together after dinner, her friends too began to engage with her on the same topic. However, they took a different line of enquiry—'Is Kamalesh asking you to leave sports?' they asked, outrage writ large on their faces. They, of course, knew Kamalesh as a peer, and knew him as a young man with a vile temper. She reluctantly agreed that there was pressure from him to quit—he did not like her to spend all day at the sports ground, clad in tight clothes, in view of other young men. Even if she went to practise surreptitiously, he would find her and his eyes would turn red and menacing; '*avanakku very varum*', 'he gets into such a rage', she said. '*Enakku indha madhiri sandai poattutu irukka mudiyaadhu ... Thaangle!*' 'I cannot keep fighting with him. I cannot take it anymore ...!', she continued, tears shining in her eyes.

In the next few days, the conflict between Rathna's friends and Kamalesh flared up dramatically. Being his classmates, some of them had bypassed Rathna and confronted him directly and he had become furious. He asked Rathna never to speak to them again. Poor Rathna was caught in a double bind—how could she stay away from her friends when they were her room-mates? How could she break up with her boyfriend when they loved each other? But her friends stood their ground—'let her pick whom she wants in her life: the ones who stood up for her, or the one who made her quit sports', they said. Kamalesh was also steadfast in his stand: 'who are your friends to tell me how to behave with my girlfriend?! Don't you dare speak to them again!'. For days together, Rathna lay on her bed, weeping into her pillow—confused and anxious, and clearly in distress. She would try to explain things on the phone to Kamalesh only to fight and argue. He would abruptly cut the call, and she would again burst into tears. Cut off from her peer group, she ate her meals alone, eyes swollen from continuous sobbing.

This episode symbolized the first of many negotiations and compromises that Rathna would eventually have to make in order to stay in a relationship with her chosen partner—one who supplied her with care and nurture, but also interpreted his care and nurture within a protectionist paradigm of saving her from other men's gaze, and the supposedly malevolent intentions of her friends. Rathna's emotional upheaval seemed to be a precursor to what she would eventually face when she tried to convince her parents about Kamalesh's suitability as a life partner, underlining her vulnerability as she negotiated between two systems of control, both claiming to have her interests at heart.

The gendered subjectivities encountered in the making of the relationship are also striking—Kamalesh's masculine subjectivity seemed to underpin both his sense of care and the sense of control that he attempted to impose. There were also similarities in the objections raised to her sports career by her parents and Kamalesh—both finding it inappropriate for a young woman to pursue, even if for different reasons. Rathna, too, was wracked by anxiety and sadness, as she attempted to negotiate between her partner and friends—arguing, resisting, and challenging, but also attempting to mollify, placate, offering her love and trying to make peace. This episode, while pushing the enquiry into intimacy as a site of disciplining and shaping into respectable adult and marriageable status, also shows the ways in which multiple intimacies result in multiple subjectivities that imbricate with one another, resulting in zones of contention and conflict. The young women are positioned in ways that make them the target of concentric frames of discipline that seem in continuum with one another. However, as I have shown in multiple case studies, this does not necessarily mean women are passive in their relationships: as Rathna's continuous arguments and negotiations show, such conditions and conflicts are the 'shifting exigencies' in which Rathna remade herself as both friend and lover. Perhaps this fight even contributed to a deepening of the couples' intimacy in the face of trial, a sense of the indissolubility of the relationship. Certainly, in the Tamil movies that my interlocutors watched, such tropes were common: lovers invariably had to face and fight many challenges before being united in marriage, and living together 'happily ever after'.

However, in the years following their graduation, the 'fight' was over even before it began. When Rathna told her businessman father about her relationship with Kamalesh, her father responded angrily and told her to break off the relationship immediately. Rathna begged him for a chance: 'Meet Kamalesh once', she told him. 'He is a Mechanical Engineer and works in a very well-known automobile firm in Chennai, and he will be a good match', she begged. However, her entreaties were only met with a cold silence from her father. The silence dragged on for months, with neither father nor daughter ready to relent from their respective positions. Unsure of whether he should share these events with his parents, Kamalesh told only his elder sister of his desire to marry Rathna. She cautioned him against moving further unless Rathna could make progress with her parents, and insisted on speaking to Rathna's parents herself to understand their stance. Rathna's father did not relent and the extended deadlock finally broke the relationship. Rathna and Kamalesh started distancing themselves from each other, as her parents relentlessly looked for a suitable groom for her. When they came across Murali's profile on a matrimonial site, they felt they had found the perfect match indeed. Apart from similar caste affiliations, he was tall, athletic, and played football for a local club in Singapore, where he had a job with an engineering firm. They could not imagine a more compatible match for their daughter whose sporting career had always caused them worry about her marriage prospects. An engagement date was fixed and a suitable *muhurtham* was chosen. Rathna said she was trying her best to put painful experiences and sweet memories behind her to begin a new life with Murali. 'I tried my best ... but nothing was in my control. Maybe it is all for the best. Murali is well settled and he loves sports', she said, as she forwarded me the invitation card for her wedding ceremony. The invitation was printed on a picture from their engagement, showing a close-up of Murali's fingers slipping a ring on Rathna's bejewelled hands. 'As the sun and moon exist together for perfect days, we have decided to make our days and lives one forever', it said, followed by a few words of invitation, deftly painting the union in romantic colours of cosmic proportions, rather than highlight the short-term 'arranged' involvement of the couple. However, what it does highlight is the strength of romantic and individualistic notions even in the most conservative of settings. Even when unions are decided by the diktats of family and community, the relationship is portrayed in ways that highlights consent and decision showing its importance in the public eye, even as it is a highly contested term in the private domain.

Conclusion

The social and political risk associated with falling in love forms the mains reason behind women's precarious positions in higher education. *Doing love* is a constant process of strategizing to overcome risk at various levels, while perfecting the traits of demure femininity, respectability, and 'good' heterosexuality. Given the disciplinary frameworks of caste, gender, and family that imbricates with the disciplinary techniques of the college management, intimate desires, subjectivities, performativity, and processes of self-making emerged as a loosely defined set of what I refer to as 'intimate aspirations', following Veena Das (2010). This term, while attempting to grapple with the nature of relationships in college characterized by flux, also provides a method of thinking through the contingencies of these aspirations, the ways in which their fulfilment was strategized, their tendency to disrupt causal thinking, and the shifting nature of gendered subjectivity and self-making.

The first part of the chapter addresses marriage aspirations. Young women from upwardly mobile groups, while negotiating the difficult conditions of access to higher education, are influenced by notions of status enhancement. Parental approval occupies a place of utmost importance in the narratives of my interlocutors—which in turn rides on status and respectability in ways which are compatible with modern careers, professions, and notions of success. This chapter, like the work of Shalini Grover (2011) and Geert De Neve (2016) on contemporary love marriage, rejects the idea that modern relationships are entirely characterized by 'individualism'. Instead, I look at how 'subjectivities of suitability' are shaped by historical events and circumstances in the wider political order (such as inter-caste rivalries). While retaining a certain modern romantic and companionate ideal, 'convincing parents' remained the cornerstone of the decisions of my young interlocutors as they negotiated their choices and desires in what were sometimes very difficult circumstances. Whether on the question of dowry or status, my young interlocutors strategized possible ways in which their parents would not have to 'lose face' because of their choices. By highlighting the role of women's agency in such a context, in crafting what I call 'subjectivities of suitability', I argue that women gauge and assess relationships for acceptability, even as they find ways to negotiate their desires. To make sense of the shifting nature of suitability and the gendered performativity involved, however, one must engage sufficiently with the students' engagement with risk and pleasure.

The second part of the chapter therefore moves into the actual modes of 'doing love'—attempting to unravel gendered subjectivities, shifting aspirations, lived experiences, and processes. Both Reshma and Rathna's narratives as well as the others in this chapter show the myriad complexities of human life, within which decisions about marriage, education, and employment are taken. Careful strategy and management of the above three factors are important in crafting the family's image in the community: a continued education can successfully hide a daughter's resistance to marriage, and forced break-ups may be assuaged through the construction of new kinds of compatibilities, even as a family may encourage 'love' in a couple deemed suitable for each other. A family's respectability is therefore very carefully constructed and young women's involvement/discontinuation of education, employment, as well as marriage emerge as important tools of the same.

6

Engineering Aspirations and Lives of Youth

Implications for Gender, Caste, and Class

Life in engineering college shows not just the investment in an educational degree, but the ways in which these have various widespread implications that are tied to models of patronage, caste networks, opportunities for employment, and everyday relationships between men and women. Thus, the book aims to provide some insight not just into the lives of youth, but the larger narratives of caste, class, and gender in which they are enmeshed, providing a distinct method of looking at the complexities of youth as a life stage. In other words, I construct 'youth' as structured by the needs and objectives of private capital, in some cases scaffolded by interests of caste/religion, collective efforts to secure upward mobility, and the formation of adult identities tied to the *childhood habitus* (Trawick 1990, 143–145).

While the book reflects case studies and insights from engineering colleges in Tamil Nadu, its relevance is much wider given the context of increased investment in privatized higher education, the unprecedented employability crisis especially among engineers, and the cultural/political milieu that dictates the grain along which youth affective relationships are politicized. This volumes, therefore, looks at the complex ways in which young people negotiate their lives in higher education, the complexities of the 'regimes of aspiration' that ensures increased access for women and other disadvantaged groups into higher education, but also creates subjectivities of gendered respectability, of being 'local' and of the 'casted' suitability of marriage partners.

Caste and its Networks

One of the key contributions of the book is to trace the emic understanding of caste as being tied to the formation of networks at various levels, from organization of funds for colleges to the reliance on caste networks for jobs, and for surveillance on everyday behaviours. Thus, even though caste is delegitimized in the public sphere and notions of purity pollution are said to be losing relevance (Jodhka 2015b), we see the reconfiguration of caste as formal and informal networks that work to shape opportunities for its members, determines the pathways of private capital, enforces the group's moral (often gendered) boundaries, and influences members' subjectivities.

Moreover, education acts as a 'vector' (Massumi 2002) that intensifies 'caste feeling' in many ways similar to the formulation of affect in transnational studies (Wise and Velayutham 2005; 2017). We see intense emotions flowing with donations for education along caste/religious lines, the 'kinning' of education institutions, and their reconfiguration as patronage projects. These projects are not only meant for the delivery of education as a 'social good', but to build the careers of 'big men' as benefactors of entire communities, establishing their dominance over different caste-based regimes of labour as well as helping them cultivate 'constituents'. These institutions are used as planks to launch political careers, or expand into other education-related businesses. Thus, the privatized avatar of higher education enables modes of engagement which intensify affect, compelling people to act in ways that reaffirm their allegiance to their local groups, even when they are outside the immediate contexts of traditional face-to-face communities. Through engineering colleges, youth are enmeshed in the politics of patronage, affirming loyalties, and reiterating caste membership; these factors have a definitive role in deciding which colleges they join, the networks of which they become a part, and the cultural politics they endorse. Thus, the private engineering college forms an affective complex that heightens affects such as pride and awe, even as it enhances feelings of shame or loss of face. This is possible because the private structure of the college allows a multiplicity of factors that shrinks the distance/proximity between college and community.

I have referred to some historical examples that show the tacit role of patronage and caste networks in creating privatized models of higher education. Along with private initiatives for particular castes' 'development', the 'politics of loyalty' in former Tamil Nadu Chief Minister Maruthur Gopalan Ramachandran's (MGR) regime ensured that leaders from groups classified as 'OBCs' and 'MBCs' were put at the helm of engineering institutions, encouraging the large-scale entry of these groups into higher education. Though this enabled a sort of 'democratization of education' with backward groups gaining entry into education, it also inaugurated opportunities for many religious- and caste-based ideological organizations to legitimately enter and occupy the space of higher education. These occupations of such positions were crucial in crafting youth subjectivities in ways they deemed fit, even as they fulfilled what was perceived as an essential role in producing the workforce for the IT economy. The lives of youth in these colleges are shaped by explicitly gendered rules and regulations, and has in some parts even strengthened traditional caste rivalries such as the ones described by Martyn Rogers (2008) in his study of masculinity in colleges in Chennai (such as between OBCs and Dalits), by drawing even greater surveillance to gendered interactions.

Kongunadu, the region where I conducted the ethnographic component of my study, has seen intense investment in private colleges since the 1980s. College managements comprise wealthy industrialists, builders, and politicians, as well as caste associations. The 'dominant caste' of the region, the Gounders, and other intermediary castes such as the Vanniyars, the Naidus, and Devangas expanded from cultivating their traditional *thottams* (agricultural gardens) to trade in cotton, and eventually into export of textiles, particularly knitwear. Under the influence of globalization, these businesses have been very lucrative and the region has seen an influx of cash, leading to further industrialization and investment in industries such as construction, trucking, and poultry. This class, which has benefited from success in industrialization, is most keen to invest in middle-class markers such as the engineering education for the next generation, and can be seen as the primary driver of the proliferation of engineering colleges—both in terms of demand and supply.

However, this model of aspiration for upward mobility has also had implications for the caste politics in the region: unemployment is rampant and engineering graduates are often at a disadvantage compared to their urban peers. As I have emphasized, various factors influence 'employability'—spoken English skills, the confidence to move in globalized settings, and soft skills. These factors converge with caste and class privilege such as English-medium education and urban 'exposure'. Students' everyday lives in the college show the reproduction of the inequalities that stem from their caste and class backgrounds.

My interlocutors, in fact, faced one of the worst employability crises during their final year (2014–2015) at CCT, when few recruiters came calling, contrary to expectations. Students were especially disappointed because they had enrolled into college expecting '100 per cent placement'. After repeated protests, a host of recruiters came, offering jobs that did not need an engineering degree such as clerical jobs in banks and teaching jobs at coaching centres. Some of these offers were never even honoured, and many students waited for months, in vain, for a letter summoning them to join, making my interlocutors suspect that recruitment had been 'staged' in the first place.

After months of waiting, many had to fall back on their local networks to get jobs in cities such as Coimbatore, or had to invest further in higher education by enrolling in MBA and MTech courses. Others were forced to depend upon family income from agriculture and industry, or began to prepare for competitive exams to enter government service. An informal survey among my close interlocutors revealed that salaries did not exceed ₹3.5 lakhs per annum (about $4300 US)—less than the stipend paid by the government for higher education, and a far cry from salaries received by employees in multinational companies. What makes the situation far more poignant is that many had accessed higher education through education loans, which they struggle to pay back, often relying on parents for financial help. This is not an isolated crisis faced solely by my interlocutors, but a widespread problem in many South Indian states, including Tamil Nadu, which have seen a spate of 'engineering suicides' as a disappointed generation fails to transition successfully to adulthood (Chowdhury 2018).

The crisis in employment also has a caste dimension, and has fanned rivalries between rural OBCs and Dalits, who are perceived to be more successful because of government quotas. Apart from day-to-day confrontations between these groups on buses and outside college, such as the ones described by Rogers (2008), the conflict also has gendered overtones: caste associations have been carrying out campaigns against intercaste marriages, especially with Dalits. A few such marriages have ended tragically: the case of Divya and Ilavarasan in 2013 is the most well-known. In other cases, Dalit young men have been killed for merely talking to young women of other castes such as the Gokulraj murder case in June 2015. Communities have bestowed public honour on those behind these deaths, even as courts have convicted them. On the one hand, these incidents show the texture of caste polarizations that exist today, especially the mobilization of the 'non-Dalit bloc' not just in Tamil Nadu, but also in places such as Haryana and Maharashatra where dominant castes such as Jats and Marathas have mobilized against the Dalits, and regularly clash with them on similar issues (Chowdhry 2007; Rao 2009). On the other, they show the gendered texture of the entry of women into higher education—under conditions of surveillance in which affective relationships are policed and politicized.

Class Conundrums

Another key contention in the book is centred on the scale of investment in middle-class identities through education, producing professions such as engineering as de facto choices. I argue that an important part of how engineering education has acquired such a status is through its visibility—striking architecture that marries the economic with the cultural, claiming to inculcate an embodied *habitus* through features such as professional trouser suit uniforms and the ease of moving around in a neoliberal landscape, even while recreating the local moralscape. These have been crucial in fashioning engineering college as a consumption choice that seeks to validate and establish a middle-class identity of being upwardly mobile and cosmopolitan but 'within limits' (Gilbertson 2018). Each private college has its own formulae as to what constitutes the perfect combination of this, and their respective rules cover a spectrum from absolutely forbidding cross-sex interaction, banning the use of mobile phones, prescribing strict uniforms, to allowing students of the opposite sex to interact in 'professional settings', and enforcing dress codes that inculcate the rudiments of what would be appropriate dressing in the future workplace. These vary by location of the college, the affiliations of the management, the perception of their 'market', and the 'needs' of the recruitment networks to which they cater. These highlight the ways in which colleges are into ensuring respectability, especially for young women, but also inculcating cosmopolitanism as a key trait through higher education—though the meanings attributed to 'cosmopolitanism' can be quite relative (Gilbertson 2018), and interpreted through a narrow moral lens. For instance, in the Tamil Nadu context, chastity is seen as a virtue unique to 'Tamil women' with emphasis on cultural tropes such as the Kannagi myth that glorifies chastity (Lakshmi 1990; Vera-Sanso 2006). In CCT, this distrust could be seen in the stereotypes around 'NRIs' (students from other parts of the country and the world), their clothing and their tendency to party, drink, and have 'loose morals'. Local students were asked to stay away from the 'corrupting influences' of foreign students, even as, in the classrooms, students are taught how to adapt to globalized milieus.

Therefore, a certain degree of cosmopolitanism is considered desirable. Amanda Gilbertson (2018) argues in her book on middle classes in Hyderabad that such investment in ideas of cosmopolitanism run deep into regimes of schooling, especially in privatized elite schooling, along with the academic pressure to perform well in various examinations. This is not only specific to the 'super schools' of Kongunadu, but are visible in much of small-town India, where coaching classes for entry into engineering and medical schools begin as early as class eight. This is not surprising considering that the investment in the middle-class child's success in white-collar employment is thought to begin with the selection of preschool for the child (Donner 2006). Moreover, the need to secure a competitive edge in streams like engineering has now acquired such an intensity that children as young as age six can take software coding classes online and toys that inculcate skills in software coding are marketed as suitable for three-year-old children. Henrike Donner (2006), through her study of early years schooling in Kolkata, draws our attention to the ways in which the need to prepare the child for success has become a key component in the construction of middle-class parenthood and adolescence.

Moreover, the academic pressure in engineering college is mostly focused on maintaining a consistent academic record (passing examinations over excellence and innovation), but more than anything else preparing and doing well in recruitment tests, hoping to make the 'cut' for a job. In such a scenario, almost all the pressure in engineering college is geared towards a certain kind of class-making—emerging articulate in English, developing good verbal and written communication and soft skills, developing a favourable worker subjectivity. These class traits are also highly valued, and are considered one of the most compelling reasons to send children to engineering college (De Neve 2011), even if job prospects are somewhat lacklustre, and most students acknowledge that they would have to probably invest even more in higher education to secure actual employment. This emergence of engineering education in private colleges as a form of basic qualification symbolizes the pressure on middle-class lives by the withdrawal of the state from education and the 'market capture' achieved by private engineering colleges in dominating the middle-class imaginary of success. The implication of this is the added burden on the middle class, in terms of expenditure on education, without the hope of expected returns, doing nothing more than contributing to the creation and sustenance of a 'mediocre bubble', though the middle class has also begun to abrogate the system.

Efforts to reinvent higher education and bring about better enrolment ratios such as the New Education Policy 2020 attempt to level the playing field: physics, chemistry, and maths are no longer mandatory for entry into engineering college. Engineering aspirants are now allowed to choose between thirteen subjects while appearing for entrance examinations; 'the idea is to encourage multi-disciplinarity, flexibility and choice in education' (Tukdeo, Singai, and Aiswarya 2020). Yet, not all thirteen subjects are offered by Indian school boards. Of the thirteen subjects listed, twelve are offered by the Central Board of Secondary Education (CBSE) but state schools offer only seven (Tukdeo, Singai, and Aiswarya 2020). With such discrepancies, the already uneven terrain of higher education will become harder to negotiate for students from state schools, government-run schools and for youth from rural and disadvantaged backgrounds.

Gendered Hopes and Limits

There are many celebratory accounts of women entering higher education in larger numbers, especially in urban areas. Many news magazines present images of emancipated young women on the cover of their higher education specials, and the numbers themselves are encouraging—girls are staying in school longer and more women are entering higher education than ever before. This has also been captured in important works such as Alice Clark's *Valued Daughters* (2016), in which she documents the lives of first-generation career women, who are changing and effecting change by stepping out of their traditional roles to pursue higher education and jobs. While these are encouraging trends, it is useful to juxtapose these celebratory accounts with the fact that women's involvement in the labour force is at an all-time low (NSO 2019–2020). How do we understand this paradox? What does it say about the women's status in the family, their roles, and functions?

According to my interlocutors, many parents desire that their daughters receive a higher education that is comparable to a son's. To an extent, such aspirations for daughters can be traced to the growth of the Information and Communication Technology industry and the visibility it commands in South Indian cities like Bengaluru, Hyderabad and Chennai. As the figures provided from CCT in Chapter 4 within this volume show, traditional fields of engineering such as Mechanical and Civil Engineering continue to be male-dominated, and retain notions of the mastery of technology as being a uniquely masculine trait. However, fields such as Information Technology (IT), Computer Science (CS), and Electronics and Communication Engineering (ECE) have seen an increase in women's enrolments because of the perception that they are lucrative, but also 'soft' and more suitable for women.

CCT's student body is about forty-three per cent women. The numbers in CCT are not comparable to all-India figures which show that engineering as a field of study continues to be male dominated (and that women's enrolment is less than thirty per cent of the total [Nair 2012; MHRD report 2016, 12]). CCT's location in Salem district, in the heart of industrial Kongunadu in South India is distinctive in shaping the 'aspirational regime' in which families invest heavily in professional education, as elaborated throughout this book. Moreover, outsourced industries such as IT have a special place in the urban imaginary as a 'unique site of display', as Carla Freeman (2000) also finds in the case of the bioinformatics industry in Cuba. This industry contains connotations of not only being suitable for women, but also as a form of 'high class' and 'respectable' labour.

Although families often invest in the education of both sons and daughters, there continue to be differences. Carol Mukhopadhyay and Susan Seymour (1994) label such a tendency among Indian families as 'patrifocality', or the preference given to men's education and employment over women's (sons over daughters and husbands over wifes). Despite what appears to be a narrowing gap between sons and daughters, and rising aspirations for daughters, the case studies in Chapters 2 and 4 in this volume show the gendered nature of aspiration as well as anxieties among parents investing in higher education. Smitha was 'allowed' to study in a college close to home, whereas admission was sought in a college reputed in the weaving community for her brother to showcase the family's new-found status in the community. Namitha was subject to various parental restrictions throughout her time in college—no outings, no trips home, and no male friends—to help maintain the family's respectability. Her father had ignored the advice of many of his well-wishers not to send his daughter to college, and any slip on Namitha's part could cause a loss of face for her father: this had produced a gendered 'affect' in her, and her education had become a status-building project between her father and her. Such restrictions were not just imposed by Namitha's parents but were commonly reiterated by many parents in CCT.

In my discussions of women in an engineering college in Salem, entry into higher education and employment for women emerge as fraught issues in the household and a cause of much anxiety to the middle class family. These anxieties also shape the organizational practices of colleges: parents are notified of their ward's attendance daily and students must conform to a dress code which emphasizes modesty, even as it aims at cultivating a 'professional' identity. Moreover, various mechanisms of surveillance are also used to ensure students conform to modes of acceptable behaviour such as not making aimless trips outside or staying out after sundown.

Though the aspiration to join the IT industry is, at one level, palpable, students and teachers are also critics of the industry, finding it 'morally dubious' and unsuitable. Families are known to actively discourage their children, especially daughters, from joining the IT industry. Concerns centre on men and women working together late into the night, travelling together, being posted in faraway client-sites, and the general lifestyles of IT professionals. The investment in engineering colleges is not always tied to the endgame of finding employment. Engineering degrees are also considered prestigious for the family, useful for obtaining 'exposure', 'soft skills', and other necessary attributes of middle-classness (Fuller and Narasimhan 2006; De Neve 2011). Engineering education, especially for a young woman, is therefore premised on securing respectability for the family and increasing her marriage prospects, as much as equipping her with higher education to increase the family's prestige. Many young women said that whether they had lifetime careers would depend upon their marital families. Looking after the household would remain their primary role; though they may eventually consider joining a family business, but chances were low that they would be 'sent out' to work. Compulsory marriage and their subordinate position in the household were a given, even as they enrolled for higher education in engineering. These were also reiterated by management and guests on formal occasions such as the Women's Day as I have elaborated in these chapters.

To this end, control over women's sexuality becomes an important concern for families throughout their adolescent years—a concern also shared by the college, which reiterates a protectionist paradigm in its rules and regulations, and is embodied in teachers' and wardens' everyday attitudes towards young women in the college. This translates into loss of equal opportunity during college years, and a gradual internalization of bias: women were habituated into bowing out of situations because they were women. I have described how women could not participate in an inter-collegiate event in Chandigarh, even though they spent many hours developing the project that was their team's entry into the competition. This is relevant not only to CCT, but has been widely reported across colleges in Tamil Nadu. Women say that they are not able to take extra classes, participate in events, or spend a few extra hours in the library because of the restrictions imposed on them at home and replicated in hostels (Suryanarayan and Venkatesh 2016). Thus, it emerges that conditions that contribute to respectability are important in determining whether women enter and stay in higher education. The responses captured in the chapter, highlighting terms such as *'mudiyaadhu'* ('not possible') or that they feel *'vekkam'* (shyness) drawing attention to themselves shows the intense affect and gendered embodied dispositions produced in women as a result of their family *habitus*, which heightens feelings of shame. Young women remade their own subjectivities in line with the rules and regulations posed by the administration and their families, even as they accessed spaces of higher education, strategized resistance, conformed while dreaming of the possibilities of how to 'enjoy' college life. Boundaries therefore emerged as shifting lines—between what is desired, possible, what can be strategized or negotiated, what is plainly impossible, and what should be accepted as it is useful to 'become' professional or 'makes sure we do not misuse our freedom'.

In situations when respectability is endangered, young women run the risk of discontinuing their education and leaving without the degree. I have pointed to the trend of 'missing girls'—young women who enroll in higher education and then discontinue their studies because they have been found endangering their family's respectability. In one case that I looked at in detail, the young woman, Sudaroli, had refused a marriage with her father's sister's son (FSZ), and had been threatened that she would not go to college again. Thus, it is widely accepted that young women's primary duty is towards their home and their families, in accepting decisions made by their families, rather than practise autonomy and strategize for their careers. These biases are also internalized by young men in the college who tend to view gender in hierarchized ways.

However, women are not passive within these regimes of control: women also strategize to break hostel rules, plan leisure trips, and cultivate intimate relationships with young men. Though these relationships are sometimes explicitly labelled 'timepass' as the detailed case study of Reshma shows, most are referred to as 'true love' relationships, i.e. relationships that have marriage and securing acceptance by parents and wider kin as an objective. Such an objective also helps secure legitimacy for the relationship in the college, where multiple relationships are viewed as moral depravity, especially among young women. Performativity by young women reiterate that 'true love' is more acceptable as it carries connotations of monogamy, of being 'true' enough to have the potential of securing wider acceptance in the community. Such notions are also informed by caste and class: thus, young OBC women in relationships with OBC men were more confident of securing acceptance for their relationships. These were characterized as relationships that would not tarnish the family's sense of *geththu* (status and prestige). College relationships also tended to follow a gendered sense of hierarchy, where young men often claimed the upper hand in making decisions about where their partners go, what they do, and with whom they speak. Even as young women attempted to negotiate these aspects of their relationships and secure more autonomy, they often found themselves in very vulnerable positions in which they were at the receiving end of emotional abuse and violence, fighting against all odds to maintain a relationship of choice.

Moreover, notions of love do not always involve consent. Young women sometimes find themselves being 'stalked' or subjected to violent crimes because they are either objects of young men's affections or have rejected such overtures. Help and support for these situations are also minimal as demonstrated in what has almost become a frequent phenomenon—murders of female employees of IT firms attributed to what the newspapers call 'love angles' (Lakshmanan 2014; Janardhanan 2016). The backgrounds of these women are eerily similar: women in their early twenties, having left their small-town homes for well-paying jobs in the metropolis, staying on their own, working late on overseas projects, and making new friends and their own choices. Responses to the murders cover the usual spectrum: companies recommend self-defence classes for their employees, while social groups call for an end to the objectification and commodification of women on screen, expressing shock that these 'ills' exist among the 'educated class'.

Researching some of the education institutions that churn out the workforce for these companies, I found myself asking: how much of the culture and attitudes prevalent in engineering colleges contribute to this culture among IT professionals? What does it really mean to contract romantic and sexual relationships of choice in these parts? The unfortunate circumstances of the women's murders suggest that similar conditions are replicated in the workplace—despite economic independence and an autonomous existence in the big city, these women hardly have access to egalitarian relationships. Instead, it suggests that the processes of modernization and globalization have strengthened patriarchal structures through narratives of romance, rather than contributing towards greater choice for women and greater flexibility in entering and terminating relationships.

However, despite entertaining hopes of marriage, few college relationships end in marriage. It was reiterated to me by my interlocutors that college relationships often followed a 'script', in which despite the intentions of getting married, young people invariably gave up these relationships after college due to various reasons, most commonly because parents did not agree to give their blessings to the relationship. Young people then settle into relationships acceptable to their parents and wider community. Despite the 'true love' and 'timepass' dichotomy created in college, there is a similarity to how these relationships ultimately play out.

This is because families are very heavily invested in projects of upward social mobility: this does not mean only monetary investment in young people's education, but the crafting of respectable identities through securing a place in a prestigious college, and eventually desirable marriage alliances. Women's education especially, I argue, is premised on maintaining respectability. Yet, young women are not passive in these matters. One cannot rule out, for instance, that women are not influenced by the caste politics in the region. As I have argued, caste plays a significant role even in deciding to which college a young person gains admission, and significantly shapes everyday life in engineering college as well. I also examined the explicit ways in which caste lines are drawn in college in gendered terms, and the various rules through which young people's lives are regulated. By highlighting the role of women's agency in such a context, in crafting and embodying what I call subjectivities of suitability and respectability, I show young women find ways of enrolling and staying in higher education even in difficult circumstances, and negotiating to fulfil their desires and ambitions.

Their everyday lives are complex: hierarchy and egalitarianism are intermeshed even in friendship, which has largely been characterized in sociological literature in egalitarian terms (Rezende 2007; Nisbett 2009). My work shows that hierarchical aspects such as caste tend to be complexly intertwined with friendships. Therefore, despite many status-levelling practices among Tamil youth (Nakassis 2010), my work shows that aspects such as surveillance and ostracization are practised even in peer groups. Students' experiences in colleges are circumscribed heavily by these aspects of friendship even as peer networks are important in enabling students to access various resources and pleasures in college life.

Overall, this ethnographic volume attempts to excavate a genealogy of the modern middle-class Indian youth today—one who places great importance on higher education to become 'employable' in the service economy, become 'respectable', and even considers himself/herself to have a 'choice' in choosing his/her intimate partner by convincing parents, as long as it is governed by certain definitions of 'suitability'. In considering the question of subject and subjectivity in the milieu of an engineering college, I look at the varying personhoods inaugurated by these processes, and the ways in which inner processes are reshaped within political and economic reforms. Caste, class, and gender become important lines of enquiry for looking at both continuities and diversities of personhood emerging from these processes, showing that transitions from youth to adulthood require investment of several resources, the ability to craft 'employable' selves as per the needs of the service economy, and continued negotiation with traditional models of being male and female. In looking at different articulations of respectability, responsibility, employability, and sexuality among my interlocutors, I attempt to explore what it means to be located within these regimes of aspiration: not just in the ways listed above, but how different men and women use college life also to court risk and adventure by strategizing to break rules, entering dangerous liaisons, and crafting themselves from their subjective experiences.

Therefore, I look at subjectivities in the engineering college as the ground in which 'a long series of historical changes and moral apparatuses coalesce—in the emergence of new kinds of public–private involvements as well as a new kind of political authority' (Biehl, Good, and Kleinman 2007, 3–4). As Biehl, Good, and Kleinman (2007) suggest, such a method enables the examination of 'the interconnections among changing modes of subjectivation and transformations of social organization, modes of production, knowledge structures, and symbolic forms' (2007, 5). It helps us go beyond an examination of the political and economic aspects of private engineering education, which has been well established (Kaul 1993; Kamat 2011; Upadhya 2016b) to a mode in which we understand historically situated differences in social sensibility; cross-cultural differences in cognition, affect, and action; and the peculiarities of each individual within the widespread culture of studying engineering in India.

Implications for Studies of Youth

These narratives, continuities, contradictions, and breaks are important in marking what is distinctive about young people's lives in South

India: even as young men and women show the aspiration for globalized careers, for upward mobility, and for training to fit into an urban middle-class milieu, anxieties plague their lives and their families. Many of these are inaugurated by complex processes such as globalization and modernization, which have sharply rendered the barriers between caste and class. Higher education, too, does not always live up to its promise. Given the situations of economic threat, it is important to understand how young people reflect on their own experiences, express aspirations, and social difference.

In this work, I have showcased the lives of youth who aspire to enter the engineering profession, the different routes they take to do so, and the different social trends that have come to be associated with the aspiration to become successful engineers. Anxieties about women's autonomy and youth sociality have ensured that colleges forbid/restrict cross-sex interaction and have incorporated gender segregation even into their architecture in the form of separate staircases and canteens for men and women. By contextualizing the proliferation of engineering colleges within the social context of Kongunadu, I also show the aspiration for engineering has even shaped certain social and political movements such as the Campaign against Inter-caste Marriage (Ananth 2012).

These campaigns tie in with other political and social movements organized by educated, unemployed men who portray themselves as self-consciously 'anti-modern' or 'neo-traditional' (Jeffrey, Jeffery, and Jeffery 2008, 15). In a wide variety of settings, young men have used such stances 'to rationalize poor occupational outcomes, inure themselves against the threat of exclusion or tap into alternative sources of respect, work and sociability' (2008, 15). This is also a common argument among political scientists and sociologists studying youth and youth political movements (Jeffrey 2010). What has been left unstudied so far is the impact of these movements on everyday spaces of youth sociality such as engineering colleges, the feelings that structure action, and the affect it produces shaping certain desires as well as intensifying feelings such as hatred or shame. While my book attempts to look at these aspects, there is scope for other studies to follow suit, especially given the ever-shifting realm of gender and its intersectionalities with caste and class. For instance, it would be important and pertinent to look at how movements such as #MeToo which sent major shockwaves across elite locations have influenced spaces such as CCT. Similarly, although I have attempted to explore facets of sexuality in this volume, there is scope for further work in the ways in which young people articulate non-heteronormative identitities and relationships, and the ways these politics trouble the idea of the campus as a heteronormative space. It would also be interesting to see the ways in which the Covid-19 pandemic has influenced sociality among students and whether the popularity of dating apps have made it more acceptable to engage in relationships on campus, providing a virtual space in which students can escape the surveillance around them to pursue their intimate aspirations.

I believe that new developments in technology could have also influenced the employability race, and it would be interesting to revisit engineering colleges to see how Artificial Intelligence and the onslaught of new technologies with enhanced communication and presentation skills are affecting the way students prepare for the job market. It would also be interesting to observe the current emphasis on entrepreneurship on campus culture, given that the current Prime Minister Modi has continuously focused on building start-up infrastructure as an important part of his political agenda for youth, in triangulation with changes to the digital economy and manufacturing from India. It is said that four new start-ups are inaugurated every hour in India, and engineering colleges form an important nodal point of the infrastructure, being the space where youth engage most actively with science and technology.

This volume began with an analogy between the Poultry Corridor in the Namakkal and engineering colleges. Almost as if taking the same analogy further, the poultry metaphor is whole-heartedly applied in the start-up terrain as colleges inaugurate hatcheries and offer N.E.S.T (Nurturing Entrepreneurs with Sustainable Technologies) funds fo nurture start-ups, along with semester breaks and multiple exit points so students can emerge as entrepreneurs when they graduate. Part of this transformation is also that the institutional big men who run the colleges also see this as an opportunity to invest in the start-ups incubated at the college, furthering their own image as well as acquiring constituents in the process. Women also have a special visibility within the start-up ecosystem, with 10% of funds being reserved especially for women-run ventures. Start-ups have also acquired a particular glamour quotient thanks to shows such as Shark Tank, and students are expected to gain skills such as pitching and budgeting as part of their college training, in addition to being able to successfully market and wield influence over audiences to win. How discrepancies in language learning, confidence and other factors influence students' success in such a field would be interesting to document, even as the term 'youth' continues to carry the burden of the nation's hopes and dreams.

Bibliography

Abraham, Itty. 1998. *The Making of the Indian Atomic Bomb: Science, Secrecy and the Postcolonial State*. New Delhi: Orient Longman.

Abraham, Janaki. 2010. 'Veiling and the Production of Gender and Space in a Town in North India: A Critique of the Public/Private Dichotomy'. *Indian Journal of Gender Studies* 17, No. 2: 191–222. https://doi.org/10.1177/097152151001700201.

Abraham, Janaki. 2011. '"Why did you send me like this?": Marriage, Matriliny and the "Providing Husband" in North Kerala, India'. *Asian Journal of Women's Studies* 17, No. 2: 32–65. https://doi.org/10.1080/12259276.2011.11666107.

Abraham, Janaki. 2014. 'Contingent Caste Endogamy and Patriarchy: Lessons for Our Understanding of Caste'. *Economic and Political Weekly* 49, No. 2: 56–65. https://www.jstor.org/stable/i24477760.

Abraham, Leena. 2001. 'Redrawing the Lakshman Rekha: Gender Differences and Cultural Constructions in Youth Sexuality in Urban India'. In *Sexual Sites and Seminal Attitudes: Sexualities, Masculinities and Culture in South Asia* 14: 133–156. https://doi.org/10.1080/00856400108723441.

Abraham, Leena. 2002. 'Bhai-behan, True Love, Time Pass: Friendship and Sexual Partnerships among Youth in an Indian Metropolis'. *Culture, Health and Sexuality* No. 3: 337–353. http://dx.doi.org/10.1080/13691050110120794.

Åhäll, Linda. 2018. 'Affect as Methodology: Feminism and the Politics of Emotion'. *International Political Sociology* 12, No. 1: 36–52.

Ahearn, Laura. 2001. *Invitations to Love: Literacy, Love Letters and Social Change in Nepal*. Ann Arbor, MI: University of Michigan Press.

Ahmed, Sara. 2004. *The Cultural Politics of Emotion*. New York: Routledge.

Ahmed, Sara. 2014. 'Afterword: Emotions and their Objects'. In *The Cultural Politics of Emotion*, 2nd ed. 204–33. Edinburgh: Edinburgh University Press.

AICTE website (n.d.). 'Dashboard'. Accessed 27 June 2020. http://www.facilities.aicte-india.org/dashboard/pages/dashboardaicte.php.

AICTE website (n.d.). 'Enrollments: Gender and Category Wise'. https://facilities.aicte-india.org/dashboard/pages/angulardashboard.php#!/graphs.

AISHE 2019–2020. 'Key Results of AISHE 2019–20', Ministry for Education, Department of Higher Education. Available at https://www.education.gov.in/sites/upload_files/mhrd/files/statistics- new/aishe_eng.pdf.

All India Council for Technical Education (AICTE). 2018. 'Model Curriculum for Undergraduate Study in Engineering and Technology'. https://www.aicteindia.org/education/model-syllabus.

Alm, Björn. 2010. 'Creations Followers, Gaining Patrons: Leadership Strategies in a Tamil Nadu Village'. In *Power and Influence in India: Bosses, Lords and Captains*, edited by Pamela Price and A. R. Ruud, 1–19. New Delhi: Routledge.

Althusser, Louis. 1971. 'Ideology and Ideological State Apparatuses (Notes Towards an Investigation)'. In *Lenin and Philosophy and Other Essays*, 79–87. New York: Verso Books.

Anandhi, S., J. Jeyaranjan, and Rajan Krishnan. 2002. 'Work, Caste and Competing Masculinities: Notes from a Tamil Village'. *Economic and Political Weekly* 37, No. 43: 4397–4406. https://www.jstor.org/stable/4412773.

Ananth, M. K. 2012. 'Educated caste Hindu youth campaign against inter-caste marriages'. *The Hindu*, 16 July 2012. Accessed 5 December 2013. https://www.thehindu.com/todays-paper/tp-national/tp-tamilnadu/educated-caste-hindu-youth-campaign-against-intercaste-marriages/article3644332.ece.

Appadurai, Arjun. 1996. *Modernity at Large: Cultural Dimensions of Globalization*. Delhi: Oxford University Press.

Arnold, David. 1974. 'The Gounders and the Congress: Political Realignment in South India, 1920–1937'. *South Asia* 4: 1–20.

Baas, Michiel. 2009. 'The IT Caste: Love and Arranged Marriages in the IT Industry of Bangalore'. *South Asia: Journal of South Asian Studies* 32, No. 2: 285–307. https://doi.org/10.1080/00856400903049531.

Badiou, Alan with Nicholas Truong, 2012. *In Praise of Love*. Trans. and edited by Peter Bush. London: Serpent's Tail.

Baker, Christopher John. 1984. *An Indian Rural Economy 1880–1955: The Tamil Nadu Countryside*. Oxford: Clarendon Press.

Bapna, Geetika. 2014. 'Contemporary Meanings of Marriage'. Unpublished PhD diss., University of Delhi.

Barnett, Marguerite Ross. 1976. *The Politics of Cultural Nationalism in South India*. Princeton, NJ: Princeton University Press.

Basu, Aparna. 1982. *Essays in the History of Indian Education*. New Delhi: Concept Publishing.

Bate, Bernard. 2009. *Tamil Oratory and the Dravidian Aesthetic*. New York: Columbia University Press.

Baviskar, Amita and Raka Ray (eds). 2011. *Elite and Everyman: The Cultural Politics of the Indian Middle Classes*. New Delhi: Routledge.

Baxi, Pratiksha. 2014. *Public Secrets of the Law: Rape Trials in India*. New Delhi: Oxford University Press.

Bayly, Christopher A. 1992. *Rulers, Townsmen, and Bazaars: North Indian Society in the Age of British Expansion 1770–1870*. Cambridge and New York: Cambridge University Press.

Bayly, Susan. 1989. *Saints, Goddesses and Kings: Muslims and Christians in South Indian Society 1700–1900*. Cambridge: Cambridge University Press.

Beck, Brenda. 1972. *Peasant Society in Konku: A Study of Right and Left Subcastes in South India*. Vancouver, BC: University of British Columbia Press.

Belliappa, Jyothsna Latha, 2013. *Gender, Class and Reflexive Modernity in India*. Palgrave Macmillan: Hampshire.

Benu, Parvathi and Amritha Satish Kumar. 2022. 'Educated, but no work: Covid spurs jobless rate among graduates'. *The Hindu Business Line*, 15 July 2022. Accessed 12 October 2022. https://www.thehindubusinessline.com/data-stories/data-focus/educated-but-no-work-covid-accelerates-unemployment-rate-among-graduates/article65643658.ece.

Beteille, Andre. 2001. 'The Indian middle class'. *The Hindu*. 5 February 2001. Accessed 14 February 2017. https://www.thehindu.com/2001/02/05/stories/05052523.htm.
Beteille, Andre. 1991. 'The Reproduction of Inequality: Occupation, caste and family'. *Contributions to Indian Sociology* 25, No. 1: 3–28. https://doi.org/10.1177/006996691025001003.
Bhagat, Rasheeda. 2012. 'More than sand castles'. *The Business Line*. 9 August 2012. Accessed 25 October 2014. http://m.thehindubusinessline.com/news/variety/more-than-sand-castles/article3742649.ece.
Bhandari, Parul. 2017. 'Pre-marital Relationships and the Family in Modern India'. *South Asia Multidisciplinary Academic Journal* 16. https://doi.org/10.4000/samaj.4379.
Bhandari, Parul. 2018. ' "Friends are our chosen family": Narratives of young middle class in Delhi'. Paper presented at the 'What's New in Kinship?' Conference, IIT-Hyderabad, 1–2 February 2018.
Biao, Xiang. 2008. 'Gender, Dowry and the Migration System of Indian Information Technology Professionals'. In *Marriage, Migration and Gender*, edited by Rajni Palriwala and Patricia Uberoi, 235–260. New Delhi: Sage.
Biehl, João, Byron J. Good, and Arthur Kleinman (eds). 2007. *Subjectivity: Ethnographic Investigations*. Berkeley, CA: University of California Press.
Blickenstaff, Jacob. 2005. 'Women and Science Careers: Leaky Pipeline or Gender Filter'. *Gender and Education* 17, No. 4: 369–386. https://doi.org/10.1080/09540250500145072.
Bourdieu, Pierre. 1977. *Outline of a Theory of Practice*. Cambridge: Cambridge University Press.
Bourdieu, Pierre. 1986 [1983]. 'The Forms of Capital'. In *Handbook for Theory and Research for the Sociology of Education*, edited by J. G. Richardson, 241–258. Westport, CT: Greenwood Press.
Bourdieu, Pierre. 1990. *The Logic of Practice*. Stanford, CA: Stanford University Press.
Brainard, Suzanne G. and Carlin, Linda. 2013. 'A Six-Year Longitudinal Study of Undergraduate Women in Engineering and Science'. *Journal of Engineering Education* 87, No. 4: 369–375. https://doi.org/10.1002/j.2168-9830.1998.tb00367.x.
Jan Breman. 1974. *Patronage and Exploitation: Changing Agrarian Relations in South Gujarat, India*. Berkeley and Los Angeles: University of California Press.
Brennan, Teresa. 2004. *The Transmission of Affect*. New York: Cornell University Press.
Brown, Megan. 2003. 'Survival at Work: flexibility and adaptability in American corporate Culture'. *Cultural Studies* 17, No. 5: 713–733. https://doi.org/10.1080/0950238032000126892.
Burton, Antoinette. 1996. 'Contesting the Zenana: The Mission to Make Lady Doctors for India 1874–1885'. *Journal of British Studies* 35, No. 3: 368–397. https://www.jstor.org/stable/175919.
Busby, Cecilia. 2000. *The Performance of Gender: An Anthropology of Everyday Life in a South Indian Fishing Village*. London and New Brunswick, NJ: Athlone Press.
Butler, Judith. 1993. *Bodies that Matter: On the Discursive Limits of Sex*. New York and London: Routledge.
Canaan, Joyce E. and Wesley Shumar (eds). 2011. *Structure and Agency in the Neoliberal University*. New York: Routledge.

Castells, Manuel. 2000. *The Rise of the Network Society* (2nd edn). Malden, MA: Wiley Blackwell Publishing.

Chakravarti, Uma. 1993. 'Conceptualising Brahminical Patriarchy in Early India: Gender, Caste, Class and State'. *Economic and Political Weekly* 28, No. 14: 579–585. https://www.jstor.org/stable/4399556.

Chanana, Karuna. 2001. *Interrogating Women's Education: Bounded Visions, Expanding Horizons*. Jaipur: Rawat Publications.

Chari, Sharad. 1997. 'Agrarian Questions in the Making of the Knitwear Industry in Tirupur, India: A Historical Geography of the Industrial Present'. In *Globalising Food: Agrarian Questions and Global Restructuring*, edited by David Goodman and Michael Watts, 58–77. London and New York: Routledge.

Chari, Sharad. 2004a. 'Provincializing Capital: The Work of an Agrarian Past in South Indian Industry'. *Comparative Studies in Society and History* 46, No. 4: 760–785. https://doi.org/10.1017/S0010417504000350.

Chari, Sharad. 2004b. *Fraternal Capital: Peasant-Workers, Self-Made Men, and Globalization in Provincial India*. Palo Alto, CA: Stanford University Press and New Delhi: Permanent Black.

Chari, Sharad. 1997. 'Agrarian Questions in the Making of the Knitwear Industry in Tirupur, India: A Historical Geography of an Industrial Present'. In *Globalizing Food: Agrarian Questions and Global Restructuring*, edited by David Goodman and Michael Watts, 58–77. London and New York: Routledge.

Chatterjee, Partha. 1989. 'The Nationalist Resolution of the Women's Question'. In *Recasting Women: Essays in Indian Colonial History*, edited by Kumkum Sangari and Suresh Vaid, 233–253. New Brunswick, NJ: Rutgers University Press.

Chopra, Radhika, Caroline Osella, and Filippo Osella (eds). 2004. *South Asian Masculinities: Contexts of Change, Sites of Continuity*. New Delhi: Kali for Women.

Chowdhry, Prem. 2007. *Contentious Marriages, Eloping Couples: Gender, Caste, and Patriarchy in Northern India*. New Delhi: Oxford University Press.Chowdhry, Prem. 2009. 'First Our Jobs, then Our Girls: The Dominant Caste Perceptions of the "Rising Dalits"'. *Modern Asian Studies* 43, No. 2: 437–449. https://doi.org/10.1017/S0026749X07003010.

Chowdhury, Shreya Roy. 2018. 'India's engineering colleges have loans to pay but no jobs—but who is clearing the debt?'. *Scroll*, 11 January 2018. Accessed on 19 January 2020. https://scroll.in/article/862623/indias-engineering-graduates-have-loans-to-pay-but-no-jobs-so-who-is-clearing-their-debt.

Chunkath, Sheela Rani and Venkatesh B. Athreya. 1997. 'Female Infanticide in Tamil Nadu—Some Evidence'. *Economic and Political Weekly* 32, No. 17: 21-25+27-28. https://www.jstor.org/stable/4405340.

Clark, Alice. 2016. *Valued Daughters: First-generation Career Women*. New Delhi: Sage.

Clark-Decés, Isabelle. 2014. *The Right Spouse: Preferential Marriages in Tamil Nadu*. Stanford, CA: Stanford University Press.

Cohen, S. 2011 [1972]. *Folk Devils and Moral Panics: The Creation of the Mods and the Rockers*. Oxford: Basil Blackwell.

Crenshaw, Kimberle. 1989. 'Demarginalizing the Intersection of Race and Sex: A Black Feminist Critique of Antidiscrimination Doctrine, Feminist Theory and Antiracist Politics'. *University of Chicago Legal Forum* 1989, No. 1: 39–-167.

1Available from: http://chicagounbound.uchicago.edu/uclf/vol1989/iss1/8?utm_source=chicagounbound.uchicago.edu%2Fuclf%2Fvol1989%2Fiss1%2F8&utm_medium=PDF&utm_campaign=PDFCoverPages.

Cross, Jamie. 2012. 'Technological Intimacy: Re-engaging with gender and technology in the local factory'. *Ethnography* 13, No. 2: 119–143. http://www.jstor.org/stable/43496441.

Cross, Jamie. 2013. 'Motivating Madhu: India's SEZs and the spirit of enterprise'. In *Enterprise Culture in Neoliberal India: Studies in Youth, Class, Work and Media*, edited by Nandini Gooptu, 124–139. Abingdon: Routledge.

Cross, Jamie. 2014. *Dream Zones: Anticipating Capitalism and Development in India*. London: Pluto.

Daily News and Analysis. 2015. 'Chennai: Engineering College's Dress Code for women sparks social media outrage'. *Daily News and Analysis*, 22 Septembe 2015. Accessed 18 August 2018. https://www.dnaindia.com/india/report-chennai-engineering-college-s-dress-code-for-women-sparks-social-media-outrage-2127592.

Dale, Roger. 1982. 'Education and the capitalist state: contributions and contradictions'. In *Cultural and Economic Reproduction in Education*, edited by Micheal W. Apple. London: Routledge & Kegan Paul.

Damodaran, Karthikeyan. 2013. 'When Reel Life Depicts Love Marriages as Unreal'. *The Hindu*, 16 July 2013. http://www.thehindu.com/todays-paper/tp-national/tp-tamilnadu/when-reel-life-depicts-love-marriagesas-unreal/article4919482.ece.

Daniel, Sam. 2012. 'PMK chief's anti-Dalit campaign sparks outrage'. *NDTV*, 4 December 2012. Accessed 5 December 2013. https://www.ndtv.com/south/pmk-chiefs-anti-dalit-campaign-sparks-outrage-506443.

Das, Veena. 1988. 'Femininity and Orientation to the Body'. In *Socialisation, Education and Women*, edited by Karuna Chanana, 193–207. Hyderabad: Orient Longman.

Das, Veena. 2010. 'Engaging the Life of the Other: Love and Everyday Life'. In *Ordinary Ethics: Anthropology, Language and Action*, edited by Michael Lambek, 376–399. New York: Fordham University Press.

Dayashankar, K. M. 2019. 'Relying on admission agents, freebies to fill engg seats'. *The Hindu*, 24 July 2019. Accessed 30 July 2020. https://www.thehindu.com/news/national/telangana/relying-on-admission-agents-freebies-to-fill-engg-seats/article28692800.ece.

De Neve, Geert. 2000. '"Patronage and 'Community'": The role of a Tamil "village" festival in the integration of a town', *Journal of the Royal Anthropological Institute* 6, No. 3: 501–519. https://doi.org/10.1111/1467-9655.00029.

De Neve, Geert. 2004. 'The workplace and the neighbourhood: locating masculinities in the South Indian textile industry'. In *South Asian Masculinities: Context of Change, Sites of Continuity*, edited by Radhika Chopra, Caroline Osella, and Filippo Osella, 60–95. New Delhi: Kali for Women.

De Neve, Geert. 2006. 'Economic Liberalisation, Class Restructuring and Social Space in Provincial South India'. In *The Meaning of the Local: Politics of Place in South India*, edited by Geert De Neve and Henrike Donner, 21–43. Abingdon: Routledge.

De Neve, Geert. 2011. "Keeping it in the family': work, education and gender hierarchies among Tiruppur's industrial capitalists'. In *Being Middle Class in India: A Way of Life*, edited by Henrike Donner, 73–99. Abingdon: Routledge.

De Neve, Geert. 2016. 'The Economies of Love: Love Marriage, Kin Support, and Aspiration in a South Indian Garment City'. *Modern Asian Studies* 50, No. 4: 1220–1249. https://doi.org/10.1017/S0026749X14000742.

De Neve, Geert. 2000. 'Patronage and "Community": the role of a Tamil "village" festival in the integration of a town'. *Journal of the Royal Anthropological Institute* 6, No. 3: 501–519. https://doi.org/10.1111/1467-9655.00029

Department of Backward Classes, Most Backward Classes and Minority Welfare, Government of Tamil Nadu'. Free Education Schemes. Available from: http://www.bcmbcmw.tn.gov.in/welfschemes.htm#Free%20Education%20Schemes%A0.

Desai, Sonalde and Amaresh Dubey, 2011. 'Caste in 21st Century India: Competing Narratives'. *Economic and Political Weekly* 46, No. 11: 40–49. http://www.jstor.org/stable/41151970.

Deshpande, Ashwini and Katherine Newman. 2007. 'Where the Path Leads: The Role of Caste in Post-University Employment Expectations'. *Economic and Political Weekly* 42, No. 41: 32–39. https://www.jstor.org/stable/i40010882.

Deshpande, Ashwini and Rajesh Ramachandran. 2017. 'Dominant or Backward: Political Economy of Demand for Quota by Jats, Patels and Marathas'. *Economic and Political Weekly* 52, No. 12: 81–92. https://www.jstor.org/stable/26695615.

Deshpande, Satish. 2003. *Contemporary India*. New Delhi: Viking Penguin.

Deshpande, Satish 2006. 'Mapping the "Middle": Issues in the Analysis of the "Non-Poor" Classes in India'. In *Contested Transformations: Changing Economies and Identities in Contemporary India*, edited by Mary John, Praveen Kumar Jha, and Surinder S. Jodhka. New Delhi: Tulika Books.

Deshpande, Satish. 2013. 'Caste and Castelessness: Towards a Biography of the General Category'. *Economic and Political Weekly* 48, No. 15: 32–39. https://www.jstor.org/stable/23527121.

Deshpande, Satish and Usha Zacharias. 2013. *Beyond Inclusion: The Practice of Equal Access in Higher Education*. New Delhi: Routledge.

Dhareshwar, Vivek and R. Srivatsan. 1996. '"Rowdy-Sheeters": An essay on Subalternity and Politics'. In *Subaltern Studies IX*, edited by Shahid Amin and Dipak Chakraborty, 201–231. Delhi: Oxford University Press.

Dickey, Sara. 1993. *Cinema and the Urban Poor in South India*. Cambridge: Cambridge University Press.

Dickey, Sara. 2013. 'Apprehensions: On gaining recognition as middle class in Madurai'. *Contributions to Indian Sociology* 47, No. 2: 217–243. https://doi.org/10.1177/0069966713482963.

Dirks, Nicholas B. 1992. 'Castes of the Mind'. *Representations* No. 37 (Special Issue: Imperial Fantasies and Postcolonial Histories. 56–78. https://doi.org/10.2307/2928654.

Dirks, Nicholas B. 1976. 'Political Authority and Structural Change in Early South Indian History'. *The Indian Economic and Social History Review* 13, No. 2: 125–157. https://doi.org/001946467601300201

Donner, Henrike. 2002. 'One's Own Marriage—Love Marriage in a Calcutta neighbourhood'. *South Asia Research* 22, No. 1: 79–94. https://doi.org/10.1177/026272800202200104.Donner, Henrike. 2006. 'Committed Mothers and Well-Adjusted Children: Privatisation, Early-Years Education, and Motherhood'. *Modern Asian Studies* 40, No. 2: 371–395. https://doi.org/10.1017/S0026749X0600196X.

Donner, Henrike. 2011. 'Gendered bodies, domestic work and perfect families: New regimes of gender and food in Bengali middle-class lifestyles'. In *Being Middle Class in India: A Way of Life*, edited by Henrike Donner, 47–72. Abingdon: Routledge.

Dumont, Louis. 1980. *Homo Heirarchichus*. New Delhi: Oxford University Press.

Eck, Diane L. 1998 [1981]. *Darsan: Seeing the Divine Image in India*. New York: Columbia University Press.

Elangovan, R. 'In Namakkal, all work and no play'. *The Hindu*, 23 May 2012. Accessed 5 October 2017. https://www.thehindu.com/news/cities/chennai/in-namakkal-all-work-no-play/article3448951.ece.

Fairclough, Norman. 2010. *Critical Discourse Analysis: The Critical Study of Language*. Harlow: Longman.

Fernandes, Leela. 2006. *India's New Middle Class: Democratic Politics in an Era of Economic Reform*. Minneapolis, MN: University of Minneapolis Press.

Freeman, Carla. 2000. 'Femininity and Flexible Labour: Fashioning Gender through Class on the Global Assembly Line'. *Critique of Anthropology* 18, No. 3: 245–262. https://doi.org/10.1177/0308275X9801800302.

Fuller, Chris. 1989. 'Misconceiving the Grain Heap: A Critique of the Indian Jajmani System'. *Money and the Morality of Exchange*, 33–63. Cambridge: Cambridge University Press.

Fuller, Chris J. 1999. *The Brahmins and Brahminical Values in Modern Tamil Nadu||, Institutions and Inequalities: Essays in Honour of André Béteille*. New Delhi: Oxford University Press.

Fuller, Chris J. and Haripriya Narasimhan. 2006. 'Engineering Colleges, "Exposure" and Information Technology'. *Economic and Political Weekly* 41, No. 3: 258–262, 288. https://www.jstor.org/stable/4417705.

Fuller, Chris J. and Haripriya Narasimhan. 2007. 'Information Technology Professionals and the New Rich Middle Class in Chennai (Madras)'. *Modern Asian Studies* 41, No. 1: 121–150. https://doi.org/10.1017/S0026749X05002325.

Fuller, Chris J. and Haripriya Narasimhan. 2008a. 'From landlords to software engineers: Migration and urbanisation among Tamil Brahmans'. *Comparative Studies in Society and History* 50, No. 1: 170–196. https://www.jstor.org/stable/27563659.

Fuller, Chris J. and Haripriya Narasimhan. 2008b. 'Companionate Marriage in India: Changing marriage systems in a middle-class Brahman subcaste'. *Journal of the Royal Anthropological Institute* 14, No. 4: 736–754. https://doi.org/10.1111/j.1467-9655.2008.00528.x.

Fuller, Chris J. and Haripriya Narasimhan. 2014. *Tamil Brahmans: The Making of a Middle-Class Caste*. Chicago, IL: University of Chicago Press.

Foucault, Michel. 1977. *Discipline and Punish: The Birth of a Prison*, trans. by Alan Sheridan. Harmondsworth: Penguin.

Foucault, Michel. 1980. *Power/Knowledge*, edited by Colin Gordon. New York: Pantheon.

Foucault, Michel. 1988. 'Technologies of the Self'. In *Technologies of the Self: Lectures at the University of Vermont, 1982*, 16–49. Amherst, MA: University of Massachusetts Press.

Foucault, Michel. 1990. *The History of Sexuality*, Volume I: *An Introduction*. New York: Vintage.

Garber, Jenny. 1991. 'Feminism and Youth Culture—From Jackie to Just Seventeen'. In *Girls and Subcultures*, edited by Angela McRobbie and Jenny Garber, 1–15. London: Macmillan.

Geetha, V. and S. V. Rajadurai. 1998. *Towards a Non-Brahmin Millennium: From Iyothee to Periyar*. Kolkata: Samya.

Gerson, J. M. and Kathy Peiss. 1985. 'Boundaries, Negotiation, Consciousness: Reconceptualizing Gender Relations'. *Social Problems* 32, No. 4: 317–331. https://doi.org/10.2307/800755.

Giddens, Anthony. 1992. *The Transformation of Intimacy: Sexuality, Love and Eroticism in Modern Societies*. Stanford, CA: Stanford University Press.

Gilbertson, Amanda. 2018. *Within the Limits: Moral Boundaries of Class and Gender in Urban India*. New Delhi: Oxford University Press.

Giriprakash, K. 2018. 'Why are Karnataka's Mutts so powerful?'. *The Hindu Businessline*, 17 April 2018. https://www.thehindubusinessline.com/news/national/why-are-karnatakas-religious-mutts-so-powerful/article23576630.ece.

Gooptu, Nandini (ed.). 2013. *Enterprise Culture in Neoliberal India: Studies in Youth, Class, Work and Media*. Abingdon: Routledge.

Gorringe, Hugo. 2012. 'Caste and Politics in Tamil Nadu'. *Seminar* 633. http://www.india-seminar.com/2012/633/633_hugo_gorringe.htm.

Gorringe, Hugo. 2018. Afterword: Gendering Caste: Honor, Patriarchy and Violence. *South Asian Multidisciplinary Journal (SAMAJ)* 19. https://journals.openedition.org/samaj/4685

Government of India. 1986. *Challenge of Education: A Policy Perspective*. New Delhi: Ministry of Education, Ministry of Human Resource Development.

Govindarajulu, Rajesh. 2015. 'Seeds of Change'. *The Hindu*, 13 February 2015. https://www.thehindu.com/features/metroplus/seeds-of-change/article6891229.ece?topicpage=true&topicId=1333.

Govindarajulu, Rajesh. 2018. 'The Pioneering Educationist—Chandrakanthi Govindarajulu (1924–2002)'. *Simplicity*, 28 August 2018. https://simplicity.in/coimbatore/english/article/798/LegaCity---The-Pioneering-Educationist---Chandrakanthi-Govindarajulu--1924---2002-.

Grover, Shalini. 2011. *Marriage, Love, Caste and Kinship Support: Lived Experience of Urban Poor in India*. New Delhi: Orient Blackswan.

Gupta, Charu. 2001. *Sexuality, Obscenity and Community: Women, Muslims and the Hindu Public in Colonial India*. London: Palgrave Macmillan.

Gupta, Charu. 2009. 'Hindu Women, Muslim Men: Love Jihad and Conversions'. *Economic and Political Weekly* 44, No. 51: 13–15. https://www.jstor.org/stable/25663907.

Hall, Stuart and Tony Jefferson (eds). 2006 [1975]. *Resistance through Rituals: Youth Subcultures in Postwar Britain*. London and New York: Routledge.

Halpern, Jake. 2013. 'Amma's Multi-faceted Empire Built on Hugs'. *The New York Times*, 25 May 2013. Accessed 14 May 2021. https://www.nytimes.com/2013/05/26/business/ammas-multifaceted-empire-built-on-hugs.html.

Hancock, Mary E. 2008. *The Politics of Heritage from Madras to Chennai*. Bloomington, IL: Indiana University Press.

Hardgrave, Robert. 1969. *The Nadars of Tamilnad: The Political Culture of a Community in Change*. Berkeley and Los Angeles, CA: University of California Press.

Harriss-White, Barbara. 2001. 'Development and Productive Deprivation: Male Patriarchal Relations in Business Families and their Implications for Women in S. India'. *QEH Working Paper Series* 65. University of Oxford.

Hart, Kimberly. 2007. 'Love by Arrangement: The Ambiguity of Spousal Choice in a Turkish Village'. *Journal of the Royal Anthropological Institute* 13, No 2: 345–362. https://doi.org/10.1111/j.1467-9655.2007.00438.x https://doi.org/10.1111/j.1467-9655.2007.00438.x.

Haynes, D. E. 1987. 'From Tribute to Philanthropy: The Politics of Gift Giving in a Western Indian City'. *The Journal of Asian Studies* 46, No. 2: 339–360. https://doi.org/10.2307/2056018.

Headrick, Daniel, R. 1988. *The Tentacles of Progress: Technology Transfer in the Age of Imperialism, 1850–1940*. New York: Oxford University Press.

Hebbar, Nandini. 2018. 'Subjectivities of Suitability: "Intimate Aspirations" in an Engineering College'. *South Asian Multidisciplinary Journal (SAMAJ)* 19. https://doi.org/10.4000/samaj.4578.

Hebbar, Nandini. 2020. 'The Unemployed Hero in Indian Tamil Cinema: Perspectives on Youth, Gender and Aspiration'. *Studies in South Asian Film and Media* 11, No. 1: 57–69. https://doi.org/10.1386/safm_00019_1.

Hebbar, Nandini N. 2017. 'Engineering Respectability: The Politics of Aspiration in an Engineering College'. *Subversions: A Journal of Emerging Research in Media and Cultural Studies*. http://subversions.tiss.edu/vol-5/nandini/.

Hebbar, Nandini and Ravinder Kaur. 2021. 'Becoming Professional, being Respectable: The Symbolic Politics of College Dressing in South India'. In *Visual and Cultural Identity Constructs of Global Youth and Young Adults*, edited by Fiona Blaikie, 195–212. Abingdon and New York: Routledge.

Heitmeyer, Carol. 2016. 'Intimate Transgressions and Communalist Narratives: Interreligious Romance in a Divided Gujarat'. *Modern Asian Studies* 50, No. 4: 1277–1297. https://www.jstor.org/stable/24734810.

Henry, Nikhila. 2013. 'Was Kamran denied justice at EFLU?'. 14 March 2013. *The Times of India*. https://timesofindia.indiatimes.com/city/hyderabad/Was-Kamran-denied-justice-at-Eflu/articleshow/18962864.cms.

Hirani, Raj Kumar (dir.). 2009. *3 Idiots*. Film. Mumbai: Vinod Chopra Films.

Hirsch, J. and H. Waldrow. 2006. *Modern Loves: The Anthropology of Romantic Courtship and Companionate Marriage*. Ann Arbor, MI: University of Michigan Press.

Hochschild, A. R. 2003 [1983]. *The Managed Heart: Commercialisation of Human Feeling*. Berkeley, CA: University of California Press.

Ilaiah, Kancha. 1996. *Why I am not a Hindu: A Sudra Critique of Hindutva Philosophy Culture and Political Economy*. Calcutta: Stree Samya.

India.com. 2016. 'NEET 2016: DMK, PMK oppose Supreme Court's Decision to hold Common Test for Medical Admissions'. *India.com*, 29 April 2016. Accessed 13 May 2021. https://www.india.com/news/india/neet-2016-dmk-pmk-oppose-supreme-courts-decision-to-hold-common-test-for-medical-admission-1146329/.

India Skills Report. 2018. 'Future Skills Future Jobs'. Gurugram, Haryana: United Nations Development Programme. Available at: https://www.undp.org/india/publications/india-skills-report-2018.

Irschick, Eugene F. 1986. *Tamil Revivalism in the 1930s*. Madras: Crea-A Publishers.

Jaffrelot, Christophe. 2003. *India's Silent Revolution: The Rise of the Low Castes in North Indian Politics*. Ranikhet: Permenant Black.
Janardhanan, Arun. 2016. 'Female Infosys Employee Hacked to Death at Chennai Railway Station'. *The Indian Express*, 27 June 2016. Accessed 17 May 2021. https://indianexpress.com/article/india/india-news-india/female-infosys-employee-hacked-to-death-in-chennai-2874289/.
Jaware, Aniket. 1998. 'The Silence of the Subaltern Student'. In *Subject to Change: Teaching Literature in the Nineties*, edited by Susie Tharu, 107–124. Hyderabad: Orient Blackswan.
Jeffrey, Craig. 2010. *Timepass: Youth, Class and the Politics of Waiting in India*. Stanford, CA: Stanford University Press.
Jeffrey, Craig, Patricia Jeffery, and Roger Jeffery. 2008. *Degrees Without Freedom? Education, Masculinities and Unemployment in North India*. Stanford, CA: Stanford University Press.
Jeffery, Patricia and Radhika Chopra (eds). 2005. *Educational Regimes in Contemporary India*. New Delhi and London: Sage.
Jeffery, Patricia, Roger Jeffery, and Craig Jeffrey. 2006. '*Parhai ka Mahaul*: An Educational Environment in Bijnor, Uttar Pradesh'. In *The Meaning of the Local: Politics of Place in South India*, edited by Geert De Neve and Henrike Donner, 116–140. Abingdon: University College London Press.
Jeffrey, Robin and Assa Doran. 2013. *Cell Phone Nation: How Mobile Phones Have Revolutionized Business, Politics and Ordinary Life in India*. Gurgaon: Hachette India.
Jodhka, Surinder. 2015a. *Caste in Contemporary India*. New Delhi: Routledge.
Jodhka, Surinder. 2015b. 'Ascriptive Hierarchies: Caste and Its Reproduction in Contemporary India'. *Current Sociology* 64, No. 2: 1–15. https://doi.org/10.1177/0011392115614784.
Jodhka, Surinder and Katherine Newman. 2007. 'In the Name of Globalization: Meritocracy, Productivity, and the Hidden Language of Caste'. *Economic and Political Weekly* 42, No. 41: 4125–4132. http://dx.doi.org/10.2307/40276546.
John, Mary. 2015. 'Intersectionality: Rejection or Critical Dialogue?'. *Economic and Political Weekly* 50, No. 33: 72–76. https://www.jstor.org/stable/24482414.
John, Mary. 2019. 'Sexual Violence 2012–2018 and #MeToo: A Touchstone for the Present'. *The India Forum*, 15 Apri 2019. https://www.theindiaforum.in/article/sexual-violence-2012-2018-and-metoo.
Kalam, Abdul A. P. J. (n.d.). 'Pledge for Students'. APJ Abdul Kalam website. Accessed 20 July 2020. http://www.abdulkalam.com/kalam/theme/jsp/oath/oath_students.jsp.
Kalpagam, U. 2008. ' "Americ Varan" Marriages among Tamil Brahmans: Preferences, Strategies and Outcomes'. In *Marriage, Migration and Gender*, edited by Rajni Palriwala and Patricia Uberoi, 98–124. New Delhi: Sage.
Kamat, Sangeeta. 2011. 'Neoliberalism, Urbanism and the Education Economy: Producing Hyderabad as a Global City'. *Discourse: Studies in the Cultural Politics of Education* 32, No. 2: 187–202. https://doi.org/10.1080/01596306.2011.565639.
Kanagaraj, Lokesh (dir.). 2017. *Maanagaram*. Film. Chennai: Potential Studios.
Kannabiran, Vasant and Kalpana Kannabiran. 1991. 'Caste and Gender: Understanding Dynamics of Power and Violence'. *Economic and Political Weekly* 26, No. 37: 2130–2133. https://www.jstor.org/stable/41626993.

Kapur, Devesh and Pratap Bhanu Mehta. 2004. 'Indian Higher Education Reform: From Half-Baked Socialism to Half-Baked Capitalism'. CID Working Paper No. 2004.108, Harvard University.

Karunya University (n.d.). 'Founders'. Accessed 17 May 2020. https://www.karunya.edu/founders.

Kaul, Rekha. 1993. *Caste, Class and Education: Politics of the Capitation Fee Phenomenon in Karnataka*. New Delhi: Sage.

Kaur, Ravinder (ed.). 2016. *Too Many Men, Too Few Women: Social Consequences of Gender Imbalance in India and China*. New Delhi: Orient Blackswan.

Kaur, Ravinder and Rajni Palriwala (eds). 2014. *Marrying in South Asia: Shifting Concepts, Changing Practices in a Globalising World*. New Delhi: Orient Blackswan.

Khan, Hamza. 2016. 'Thousands of liquor bottles, used condoms found daily on JNU campus: says BJP MP'. *Indian Express*, 23 February 2019. Accessed 17 May 2020. https://indianexpress.com/article/india/india-news-india/thousands-of-liquor-bottles-used-condoms-found-on-jnu-campus-every-day-bjp-mla/.

Kimmel, Michael S. 2001. 'Masculinity as Homophobia: Fear, Shame and Silence in the Construction of Gender Identity'. In *The Masculinities Reader*, edited by Stephen M. Whitehead and Frank J. Barrett, 266–287. Cambridge: Polity Press.

Kleinman, Arthur and Erin Fitz-Henry. 2007. 'The Experiential Basis of Subjectivity: How Individuals Change in Contexts of Societal Transformation'. In *Subjectivity: Ethnographic Investigations*, edited by João Biehl, Byron J. Good, and Arthur Kleinman, 52–65. Berkeley, CA: University of California Press.

Kolappan, B. 2012. 'Ramadoss consolidates intermediate caste groups against Dalits'. *The Hindu*, 2 December 2012. https://www.thehindu.com/news/national/tamil-nadu/Ramadoss-consolidates-intermediate-caste-groups-against-Dalits/article12432099.ece.

Kongu College of Engineering (n.d.). 'About Kongu College of Engineering'. Accessed 20 June 2016. https://www.kongu.ac.in/pages/aboutkec.php.

Krishna, Anirudh and Vijay Brihmadesam. 2006. 'What Does It Take to Become a Software Professional?'. *Economic and Political Weekly* 41, No. 30: 3307–3314. https://www.epw.in/journal/2006/30/special-articles/what-does-it-take-become-software-professional.html#.

Krishnamurthy, Mathangi. 2018. *1-800-Worlds: The Making of the Indian Call Centre Economy*. New Delhi: Oxford University Press.

Krishnan, Sneha. 2014. *Making Ladies of Girls: Middle-Class Women and Pleasure in Urban India*. PhD diss., University of Oxford.

Krishnan, Sneha. 2015. 'Agency, Intimacy and Rape Jokes: an ethnographic study of young women and sexual risk in Chennai'. *Journal of the Royal Anthropological Institute* 22, No 1: 67–83. https://doi.org/10.1111/1467-9655.12334.

Krishnan, Sneha. 2017. 'Anxious Notes on College Life: The Gossipy Journals of Elanor MacDougall'. *Journal of the Royal Asiatic Society* 27, No. 4: 575–589. https://doi.org/10.1017/S1356186317000293.

Krishnan, Sneha. 2018. 'Style-ish Girls and Local Boys: Young Women and Fashion in Chennai'. In *Styling South Asian Youth Cultures*, edited by Lipi Begum, Rohit K. Dasgupta, and Reina Lewis, 49–64. London: Bloomsbury.

Kshatriyan, Mohan (dir.). 2020. *Draupathi*. Film. Chennai: GM Film Corporation & 7G Films.

Kumar, Nita. 2011. 'The Middle-Class Child: Ruminations on Failure'. In *Elite and Everyman: The Cultural Politics of the Indian Middle Classes*, edited by Raka Ray and Amita Baviskar, 220–245. New Delhi: Routledge.

Laidlaw, James. 1995. *Riches and Renunciation: Religion, Economy and Society among the Jains*. Clarendon: Oxford University Press.

Lakshmanan, K. V. 2014. 'Chennai: "Love Angle" probed in TCS employee death'. *Hindustan Times*, 22 February 2014. Accessed 23 February 2014. https://www.hindustantimes.com/india/chennai-love-angle-probed-in-tcs-employee-death/story-sR95vdnVP15u566P3qyIpL.html.

Lakshmi, C. S. 1990. 'Mother, Mother-Politics and Mother-Community in Tamil Nadu'. *Economic and Political Weekly* 25, No. 42–43: 72–83. https://www.jstor.org/stable/4396895.

Lakshmi, C. S. 2008. 'A Good Woman, A Very Good Woman'. In *Tamil Cinema: The Cultural Politics of India's Other Film Industry*, 17–28. Abingdon: Routledge.

Lambert, Helen. 2000. 'Sentiment and Substance in North Indian Forms of Relatedness'. In *Cultures of Relatedness: New Approaches to the Study of Kinship*, edited by Janet Carsten, 73–89. Cambridge: Cambridge University Press.

Levinson, Bradley, Douglas Foley, and Dorothy Holland (eds). 1996. 'The Cultural Production of the Educated Person: An Introduction'. In *The Cultural Production of the Educated Person*, edited by Bradley Levinson, Douglas Foley, and Dorothy Holland, 1–56. Albany, NY: State University of New York Press.

Lukose, Ritty. 2009. *Liberalization's Children: Gender, Youth and Consumer Citizenship in Globalizing India*. New Delhi: Orient Blackswan.

Lukyx, Aurolyn. 1996. 'From Indos to Profesionales: Stereotypes and Student Resistance in Bolivian Teacher Training'. In *Schooling and Cultural Production of the Educated Person: Critical Ethnographies*, edited by Bradley A. Levinson, Douglas E. Foley, and Dorothy C. Holland, 239–272. Albany, NY: State University of New York Press.

Luttrell, Wendy. 1996. 'Becoming somebody in and against School: Toward a Psycho-cultural Theory of the Gender and Making Process'. In *Schooling and the Cultural Production of the Educated Person: Critical Ethnographies*, edited by Bradley Levinson, Douglas Foley, and Dorothy Holland, 93–117. Albany, NY: State University of New York Press.

Lynch, Katherine. 1990. 'Reproduction: The role of cultural factors and education mediators'. *British Journal of Sociology* 11, No. 1: 3–20. https://doi.org/10.1080/0142569900110101.

Madras Institute of Technology website (n.d.). Accessed 12 March 2018. http://www.mitindia.edu/en/home/objective/125-mitindia/home.

Malish, C. M. and P. Ilavarasan. 2016. 'Higher education, reservation and scheduled castes: exploring institutional *habitus* of professional engineering colleges in Kerala'. *Higher Education* 72, No. 5: 603–617. https://doi.org/10.1007/s10734-015-9966-7.

Massumi, Brian. 2002. *Movement, Affect, Sensation: Parables for the Virtual*. Durham, NC: Duke University Press.

McGuire, Meredith Lindsay. 2013. 'The embodiment of Professionalism: Personality Development Regimes in New Delhi'. In *Enterprise Culture in Neoliberal India: Studies in Youth, Class, Work and Media*, edited by Nandini Gooptu, 109–124. Abingdon: Routledge.

McMillin, Divya C. 2006. 'Outsourcing Identities'. *Economic and Political Weekly* 41, No. 3: 235–241. https://www.jstor.org/stable/4417702.

Menon, Jaya and Saravanan. 2011. 'Salem women abort on predictions'. *The Times of India*, 4 September 2011. Accessed 5 December 2014. https://timesofindia.indiatimes.com/city/chennai/Salem-women-abort-on-predictions/articleshow/10098989.cms.

Menon, Nivedita. 2015. 'Is Feminism about "Women?". A Critical View on Intersectionality from India'. *Economic and Political Weekly* 50, No. 17: 37–44. https://www.jstor.org/stable/24481823.

Menon, Nivedita and Aditya Nigam. 2007. *Power and Contestation: India since 1989*. London and New York: Zed Books.

MHRD. 2016. *All India Survey in Higher Education 2015–16*. New Delhi: Department of Higher Education.

Mines, Mattison. 1973. 'Social Stratification among Muslim Tamils in Tamilnadu, South India'. *Caste and Social Stratification Among the Muslims*. 61–71. New Delhi: Manohar Press.

Mines, Mattison. 1984. *The Warrior Merchants: Textiles, trade and territory in south India*. Cambridge: Cambridge University Press.

Mines, Mattison. 1996. *Public Faces, Private Lives*. Berkeley, CA: University of California Press.

Mines, Mattison. 2014 'The Political Economy of Pre-eminence and State in Chennai'. In *Patronage as Politics*, edited by Anastasia Piliavsky, 39–66. Cambridge: Cambridge University Press.

Mines, Mattison and Vijayalakshmi Gourishankar. 1990. 'Leadership and Individuality in South Asia: The Case of the South Indian Big-man'. *The Journal for Asian Studies* 49, No. 4: 761–786. https://doi.org/10.2307/2058235.

Mody, Perveez. 2008. *The Intimate State: Love Marriage and the Law in Delhi*. New Delhi: Routledge.

Mukhopadhyay, Carol. 1994. 'Linking Family Structure and Indian Women's Participation in Science and Engineering'. In *Women, Education and Family Structure in India*, edited by Carol Mukhopadhyay and Susan Seymour, 103–32. Denver, CO: West View.

Mukhopadhyay, Carol. 2004. 'A Feminist Cognitive Anthropology: The Case of Women and Mathematics'. *Ethos* 32, No. 4: 458–492. https://doi.org/10.1525/eth.2004.32.4.458.

Mukhopadhyay, Carol. 2009. 'How Exportable are Western Theories of Gendered Science? A Cautionary Word'. In *Women and Science in India: A Reader*, edited by Neelam Kumar, 137–177. New Delhi: Oxford University Press.

Mukhopadhyay, Carol and Susan Seymour (eds). 1994. *Women, Education and Family Structure in India*. Boulder, CO: West View.

Muruganandham, T. 2018. 'Come March, new Tamil dictionaries are likely to hit the stands'. *The New Indian Express*, 26 December 2018. https://www.newindianexpress.com/states/tamil-nadu/2018/dec/26/come-march-new-tamil-dictionaries-are-likely-to-hit-the-stands 1916620.html.

Nair, Janaki. 2018. 'An Inclusive Saffron'. *The Indian Express*, 24 January 2018. Accessed 22 May 2022. http://indianexpress.com/article/opinion/columns/an-inclusivesaffron-yogi-adityanath-karnataka-elections-bjp-hindutva-secularism 5036657/&hl=en-IN&tg=91&tk=17455671453171122986.

Nair, Sreelekha. 2012. 'Women in Indian Engineering: A Preliminary Analysis of Data from the Graduate Level in the Engineering Education Field from Kerala and Rajasthan'. Occasional Paper No. 58. New Delhi: Centre for Women's Development Studies. Available from: http://www.cwds.ac.in/OCPaper/OccasionalPaper58.pdf.

Nakassis, Constantin. 2010. 'Youth and Status in Tamil Nadu, India'. PhD diss., University of Pennsylvania.

Nakassis, Constantin. 2013. 'Youth masculinity, "style" and the peer group'. *Contributions to Indian Sociology* 47, No. 2: 245–269. https://doi.org/10.1177/0069966713482982.

Nakassis, Constantin. 2016. *Doing Style: Youth and Mass Media in South India.* Hyderabad: Orient Blackswan.

Nanda, Prashant K. 2016. 'AICTE reduces land requirement for opening engineering colleges'. *The Mint*, 27 January 2016. Accessed 18 July 2017. https://www.livemint.com/Home-Page/y1l07pcehZ0EKmuiRdloEK/Land-AICTE.html.

Natarajan, Swaminathan. 2010. 'Indian Dalits find no refuge from caste in Christianity'. *BBC*, 14 September 2010. Accessed 14 July 2020. https://www.bbc.com/news/world-south-asia-11229170.

Natrajan, Balmurli. 2012. *The Culturalization of Caste in India: Identity and Inequality in a Multicultural Age.* Abingdon: Routledge.

Nisbett, Nicholas. 2007. 'Friendship, Consumption and Morality: Practicing Identity and Negotiating Hierarchy in Middleclass Bangalore'. *Journal of the Royal Anthropological Institute* 13, No. 4: 935–950. https://doi.org/10.1111/j.1467-9655.2007.00465.x.

Nisbett, Nicholas. 2009. *Growing up in the Knowledge Society: Living the IT Dream in Bangalore.* New Delhi: Routledge.

Nisbett, Nicholas. 2013. 'Youth and the Practice of IT Enterprise: narratives of the knowledge society and the creation of new subjectivities amongst Bangalore's IT aspirants'. In *Enterprise Culture in Neoliberal India: Studies in youth, class, work and media*, edited by Nandini Gooptu, 175–189. Abingdon: Routledge.

NSO Survey. 2019–2020. 'Women Participation in Workforce'. *Periodic Labour Force Survey*, Ministry of Labour & Employment. New Delhi: Public Information and Broadcasting.Nurullah, Abdullah. 'MGR's education reforms cemented TN's tech base'. *Times of India*. Blog. 31 March 2016. Accessed 16 May 2016. https://timesofindia.indiatimes.com/blogs/tracking-indian-communities/mgrs-education-reforms-cemented-tns-tech-base/.

Odile, Henry and Mathieu Ferry. 2017. 'When Cracking the JEE is not enough'. *South Asian Multidisciplinary Academic Journal* 17. Trans. by Renuka George. https://doi.org/10.4000/samaj.4291.

Ong, Aihwa. 2006. *Neoliberalism as Exception: Mutations in Citizenship and Sovereignty.* Durham, NC: Duke University Press.

Ørberg, William Jacob. 2017. 'Uncomfortable encounters between elite and "shadow education" in India—Indian Institutes of Technology and the Joint Entrance Examination coaching industry'. *Higher Education* 76, No. 1: 129–144. https://doi.org/10.1007/s10734-017-0202-5.

Orsini, Francesca. 2006. *Love in South Asia: A Cultural History.* Cambridge: Cambridge University Press.

Osella, Caroline and Filippo Osella. 1998. 'Friendship and Flirting: Micro-politics in Kerala'. *Journal of the Royal Anthropological Institute* 4, No. 2: 189–206. https://doi.org/10.2307/3034499.

Osella, Caroline and Filippo Osella. 2004. 'Young Malayali men and their movie heroes'. In *South Asian Masculinities: Context of Change, Sites of Continuity*, edited by Radhika Chopra, Caroline Osella, and Filippo Osella, 224–263. New Delhi: Kali for Women.

Osella, Caroline and Filippo Osella. 2006. *Men and Masculinities in South India*. London: Anthem Press.

Pal, Joyojeet. 2019. 'Women as Software Engineers in Indian Tamil Cinema'. In *Cracking the Digital Ceiling: Women in Computing around the world*, edited by Carol Frieze and Jeria L. Quesenberg, 290–298. Cambridge: University of Cambridge. https://doi.org/10.1017/9781108609081.

Pandian, M. S. S. 1992. *The Image Trap: M. G. Ramachandran in Film and Politics*. New Delhi: Sage.

Pandian, M. S. S. 2000. 'Dalit Assertion in Tamil Nadu: An Exploratory Note'. *Journal of Indian School of Political Economy* 12: 501–517.

Parry, Jonathan. 1986. 'The Gift, the Indian Gift, and the "Indian Gift"'. *Man*, New Series 21, No. 3: 453–473. https://doi.org/10.2307/2803096.

Patel, Tulsi. 2007 *Sex-selective Abortion in India: Gender, Society and New Reproductive Technologies*. New Delhi: Sage.

Peale, Norman Vincent. 1992 [1954]. *The Power of Positive Thinking for Young People*. London: Vermilion.

Phadke, Shilpa. 2007. 'Dangerous Liasions: Women and Men: Risk and reputation in Mumbai'. *Economic and Political Weekly* 42, No. 17: 1510–1518. https://www.jstor.org/stable/4419517.

Phadke, Shilpa, Sarmeera Khan, and Shilpa Ranade. 2011. *Why Loiter? Women and Risk on Mumbai Streets*. New Delhi: Penguin.

Piliavsky, Anastasia. 2014. 'Introduction'. In *Patronage as Politics in South Asia*, edited by Anastasia Piliavsky, 1–36. Cambridge: Cambridge University Press.

Pinney, Christopher. 1997. *Camera Indica: The Social Life of Indian Photographs*. London: Reaktion Books.

Pinto, Ambrose. 1994. 'Karnataka—Politics of Capitation Fee Colleges'. *Economic and Political Weekly* 29, No. 1–2: 31–33. https://www.jstor.org/stable/4400626.

Ponniah, Ujithra. 2017. 'Reproducing Elite Lives: Women in Aggarwal Family Businesses'. *South Asian Multidisciplinary Journal (SAMAJ)* 15.

Poonam, Snigdha. 2018. *Dreamers: How Young Indians are Changing the World*. Boston, MA: Harvard University Press.

Poovey, Mary. 1984. *The Proper Lady and the Woman Writer: Ideology as Style in the Works of Wollstonecraft, Mary Shelley, and Jane Austen*. Chicago, IL: University of Chicago.

Price, Pamela. 1989. 'Kingly Models in Indian Political Behaviour: Culture as Medium of History'. *Asian Survey* 29, No. 6: 559–572. https://doi.org/10.2307/2644752.

Price, Pamela and A. E. Ruud. 2010. *Power and Influence in India: Bosses, Lords and Captains*. New Delhi: Routledge.

Radhakrishnan, Smitha. 2011. *Appropriately Indian: Gender and Culture in a New Transnational Class*. Durham, NC: Duke University Press.

Raheja, Gloria and Ann Grodzins Gold. 1996. *Listen to the Heron's Words*. Delhi: Oxford University Press.
Rajan, R. V. 2014. 'Our Own MIT: Rajam's Dream'. *Madras Musings* 23, No. 23: 16–31 March 2014. Accessed 17 February 2018. https://madrasmusings.com/Vol%20 23%20No%2023/our-own-mit.html.
Rajan, R. V. 2015. '"Vallal Chettiar"—the "Socialist Capitalist"'. *Madras Musings* 25, No. 16: 1–15 December 2015. Accessed 17 February 2018. http://www.madras musings.com/vol-25-no-16/vallal-chettiar-the-socialist-capitalist/.
Rajanayagam, S. 2015. *Popular Cinema and Politics in South India: Reimagining MGR and Rajinikanth*. New Delhi: Routledge.
Ram (dir), 2007. *Kattradhu Thamizh*. Chennai: MR Film Productions.
Ram, Arun. 2015. 'No Jeans, Skirts or Red Shirts. What Next?'. *Daily News and Analysis*, 24 September 2015. https://www.dnaindia.com/india/report-no-jeans-skirts-or-red-shirts-what-next-4433.
Ram, Kalpana. 1991. *Mukkuvar Women: Gender, Hegemony and Capitalist Transformation in a South Indian Fishing Community*. Sydney: Allen and Unwin.
Ram, Kalpana. 2013. *Fertile Disorder: Spirit Possession and its Provocation of the Modern*. Honolulu: University of Hawaii Press.Ramnath, Aparajith. 2017. *The Birth of an Indian Profession: Engineers, Industry and the State 1900–47*. New Delhi: Oxford University Press.
Rao, Anupama. 2009. *The Caste Question: Dalits and the Politics of Modern India*. Ranikhet: Permanent Black.
Ravishankar, Sandhya. 2015. 'Love, Caste and Fury: How a small-time crook came to symbolise Gounder Pride'. *Scroll*, 28 October 2015. Accessed 25 November 2016. http://scroll.in/article/762879/love-caste-and-fury-in-tamil-nadu-how-a-small-time-crook-came-to-symbolise-gounder-pride.
Rezende, Claudia Barcellos. 2007. 'Gifts of Food: Sociability and Friendship among English Middle Class People'. *Vibrant* 4, No. 2: 5–26. https://www.redalyc.org/artic ulo.oa?id=406941904001.
Richard, Analiese and Daromir Rudnyckyj. 2009. 'Economies of Affect'. *Journal of the Royal Anthropological Institute* 15, No. 1: 57–77. http://www.jstor.org/stable/ 20527634.
Rogers, Martyn. 2008. '"Modernity", "Authenticity" and Ambivalence: Subaltern Masculinities in a South Indian College Campus'. *Journal of the Royal Anthropological Institute* 14, No. 1: 79–95. https://www.jstor.org/stable/20203585.
Roohi, Sanam. 2016. 'Giving back: Diaspora Philanthropy and the transnationalization of caste in Guntur (India)'. Unpublished PhD diss., University of Amsterdam. https://pure.uva.nl/ws/files/2781636/178721_Roohi_Thesis_complete_.pdf.
Roy, Srirupa. 2007. *Beyond Belief: India and the Politics of Postcolonial Nationalism*. Ranikhet: Permanent Black.
Rudner, David. 1994. *Caste and Capitalism in Colonial India*. Berkeley, CA: University of California Press.
Rudnyckyj, Daniel. 2010. *Spiritual Economies: Islam, Globalization and the Afterlife of Development*. New York: Cornell University Press.
Säävälä, Minna. 2012. *Middle-Class Moralities: Everyday struggle over belonging and prestige in India*. Hyderabad: Orient Blackswan.

Saith, Ashwini and M. Vijayabaskar. 2005. *ICTs and Indian Economic Development*. New Delhi: Sage.
SAKSHAM Report. 2013. *Measures for Ensuring the Safety of Women and Programmes for Gender Sensitization on Campuses*. New Delhi: University Grants Commission. https://saksham.ugc.ac.in/content/downloads/SAKSHAM-BOOK.pdf.
Sancho, David. 2012. 'The Year That Can Break or Make You: The Politics of Secondary Schooling, Youth and Class in Urban Kerala, South India'. PhD diss., University of Sussex.
Sancho, David. 2013. 'Aspirational Regimes: parental educational practice and the new Indian youth discourse'. In *Enterprise Culture in Neoliberal India: Studies in youth, class, work and media*, edited by Nandini Gooptu, 160–175. Abingdon: Routledge.
Sathyabama Institute of Science and Technology website (n.d.). 'Founder Chancellor'. https://www.sathyabama.ac.in/about-us/founder-chancellor Accessed 17 June 2018.
Sen, Amartya. 1990. 'More Than a Million Women are Missing'. *The New York Review of Books*, 20 December 1990. Accessed 10 May 2018. https://www.nybooks.com/articles/1990/12/20/more-than-100-million-women-are-missing/.
Sen, Jahnavi. 2016. 'Rohith Vemula's Family, Peers Cry Foul at Smriti Irani's Claims'. *The Wire*, 26 February 2016. Accessed 18 June 2018. https://thewire.in/education/rohith-vemulas-family-peers-cry-foul-at-smriti-iranis-claims.
Seymour, Elaine and Nancy Hewitt. 1997. *Talking about Leaving: Why Undergraduates Leave the Sciences*. Boulder, CO: Westview.
Seymour, Susan. 1995. *Family Structure, Marriage, Caste and Class, and Women's Education: Exploring the Linkages in an Indian Town*.
Sharma, Mohit. 2021. 'Over 550 engineering colleges affiliated with AICTE shut shop in last 6 years'. *India Today*, 21 September 2021. Accessed 19 June 2018. https://www.indiatoday.in/education-today/news/story/over-550-engineering-colleges-affiliated-with-aicte-shut-shop-in-last-6-years-1855266-2021-09-21.
Shklar, Judith. 1991. *American Citizenship: Quests for Inclusion*. Boston: Harvard University Press.
Sidel, Mark. 2002. Review Article on Philanthropy in South Asia. 260–272. Voluntas.
Singer, Milton B. 1972. *When a great tradition modernizes: Anthropological Approach to Indian Civilization*. London: Pall Mall Press.
Singh, K. S. 1994. 'Devanga'. *People of India* No. 36, 116–124. Mysore: Anthropological Survey of India.
Singh, K.S. 1997. 'Vanniyar'. People of India No. 40. 1562-65. Mysore: Anthropological Survey of India.
Singh, Preeti and Anu Pandey. 2005. 'Women in Call Centres'. *Economic and Political Weekly* 40, No. 1: 684–688. http://www.jstor.org/stable/4416207.
Sivapriyan, E. T. B. 2019. '52% engineering seats vacant in Tamil Nadu'. *Deccan Herald*, 4 August 2019. Accessed 27 July 2020. https://www.deccanherald.com/india/52-engineering-seats-vacant-in-tamil-nadu-751988.html.
Skaria, Ajay. 2002. 'Gandhi's Politics: Liberalism and the Question of the Ashram'. The South Atlantic Quarterly 101 No. 4. 955–986.
Skeggs, Beverly. 1997. *Formations of Class and Gender: Becoming Respectable*. London: Sage.

Sridevi, S. 2014. 'Chettiar Women and Property Rights'. *Local banking and material culture amongst the Nattukottai Chettiars of Tamil Nadu*. Unpublished PhD Diss., Jawaharlal Nehru University.

Sridharan, E. 2008. 'The Political Economy of the Middle-Classes in Liberalising India'. Working Paper No. 48, Institute of South Asian Studies. https://www.isas.nus.edu.sg/wp-content/uploads/media/isas_papers/Working%20Paper%2049%20-%20Email%20-%20Political%20Economy%20of%20the%20Middle%20Classes%20in%20Liberalising%20India.pdf

Srinivas, M. N. 1984. *Some Reflections on Dowry*. The Centre for Women's Development Studies. New Delhi: Oxford University Press. https://www.yumpu.com/en/document/view/36487903/some-reflections-on-dowry-womens-studies-portal.

Srinivasan, Sharada. 2009. 'Between Daughter Deficit and Development Deficit: Situation of Unmarried Men in a South Indian Community'. *Economic and Political Weekly* 44, No. 3: 56–63. https://dx.doi.org/10.2139/ssrn.2562711.

Srivastava, Sanjay. 1996. 'The Garden of Rational Delights: The Nation as Experiment, Science as Masculinity'. *Social Analysis: The International Journal of Anthropology* No. 39: 119–148. https://www.jstor.org/stable/23171753.

Srivastava, Sanjay. 2006. *Passionate Modernity*. New Delhi: Routledge.

Subrahmanyan, Lalita. 1998. *Women Scientists in the Third World: The Indian Experience*. New Delhi: Sage.

Subramanian, Ajantha. 2015a. 'The Indian Institutes of Technology and the Social Life of Caste'. *Comparative Studies in Society and History* 57, No. 2: 291–322.

Subramanian, Ajantha. 2015b. 'Recovering Caste Privilege: The Politics of Meritocracy at the Indian Institutes of Technology'. In *Subaltern Politics* edited by Alf Gunvald Nilsen and Srila Roy, 76–100. Oxford: Oxford University Press. https://doi.org/10.1093/acprof:oso/9780199457557.003.0004.

Subramanian, Hariharan. 2008. 'Dr Jeppiaar'. Coastal Waves and Little Drops. Wordpress. 13 October 2008. Accessed 27 July 2017. https://coastalwaves.wordpress.com/2008/10/13/dr-jeppiaar/.

Subramanian, Narendra. 1999. *Ethnicity and Populist Mobilization: Political Parties, Citizens and Democracy in South India*. Delhi: Oxford University Press.

Sukumar, N. 2013. 'Quota's Children: The perils of getting educated', *Beyond Inclusion: The Practice of Equal Access in Indian Higher Education*. Edited by Satish Deshpande and Usha Zacharias. New Delhi: Routledge

Sundar, Pushpa. 2013. *Business and Community: The Story of Corporate Social Responsibility in India*. New Delhi: Sage.

Suryanarayanan, Bhargavi and Vandana Venkatesh. 2016. *Report on Conditions of Female Hostel Residents in Tamilnadu Colleges*. Unpublished: Personal Communication.

Sushma, U. N. 2012. 'Tamil Nadu colleges all set to tap solar energy', *The Times of India*, 5 November 2012. Accessed 25 October 2015. https://timesofindia.indiatimes.com/home/education/news/Tamil-Nadu-colleges-all-set-to-tap-solar-energy/articleshow/17095127.cms.

Swaminathan, Padmini. 1992. 'Technical Education and Industrial Development in Madras Presidency: Illusions of a Policy in the Making'. *Economic and Political Weekly* 27, No. 30: 1611–1622. https://www.epw.in/journal/1992/30/special-articles/technical-education-and-industrial-development-madras-presidency.

Tarlo, Emma. 1996. *Clothing Matters: Dress and Identity in India*. Chicago, IL: University of Chicago Press.

Thapan, Meenakshi. 2004. 'Embodiment and Identity in contemporary society: Femina and the "new" Indian woman'. *Contributions to Indian Sociology* 38, No. 3: 411–444. https://doi.org/10.1177/006996670403800305.

Thapan, Meenakshi. 2006. ' "Docile" Bodies, "Good" Citizens or "Agential" Subjects? Pedagogy and Citizenship in Contemporary Society'. *Economic and Political Weekly* 41, No. 39: 4195–4203. http://www.jstor.org/stable/4418762.

Tharaney, V. and Deepika Upadhyaya. 2014. 'The Burgeoning Field of Edupreneurship: A Literature Review'. *Pacific Business Review International* 7, No. 6: 69–79.

Tharu, Susie. 1998. 'Government, Binding and Unbinding: Alienation and the Subject of Literature'. In *Subject to Change: Teaching Literature in the Nineties* edited by Susie Tharu, 1–32. Hyderabad: Orient Blackswan.

Thatholu, Sidartha (dir.). 2019. '*Caste Feeling Song*'. *Amma Rajalu Kadapa Biddalu*. Film. Hyderabad: Tiger Company Production & Ajay Mysore Production. Available from: https://www.youtube.com/watch?v=JykSPfmaZRc. Last Accessed: 12 April, 2020.

The Economic Times. 2017. '60 per cent of graduate engineers unemployed', 18 March 2017. Accessed 14 August 2017. http://economictimes.indiatimes.com/jobs/60-per-cent-of-engineering-graduates-unemployed/articleshow/57700844.cms.

The Hindu. 2017. 'DMK, PMK oppose NEET', 10 April 2017. https://www.thehindu.com/todays-paper/tp-national/tp-tamilnadu/PMK-opposes-move-to-make-JEE-mandatory/article17015971.ece.

The Newsminute. 2016. 'Prominent Educationist Jeppiaar dies in Chennai', 18 June 2016. Accessed 24 July 2016. https://www.thenewsminute.com/tamil-nadu/prominent-educationist-jeppiaar-dies-chennai-45088.

The Times of India. 2008. 'Educationists among city's top landlords', 18 April 2008. Accessed 15 May 2016. http://timesofindia.indiatimes.com/city/chennai/Educationists-among-citys-top-landlords/articleshow/2960902.cms.

The Times of India. 2012. 'Now, caste outfits vow to fight inter-caste marriage', 13 October 2012. Last Accessed: 20 August 2023. https://timesofindia.indiatimes.com/city/coimbatore/now-caste-outfit-vows-to-fight-inter-caste-marriages/articleshow/16790454.cms.

The Times of India. 2012. '100 Colleges under the scanner for building violations in Chennai', 4 December 2012. https://timesofindia.indiatimes.com/city/chennai/100-colleges-under-scanner-for-building-violations-in-Chennai/articleshow/17473127.cms.

The Times of India. 2016. 'Three Girls Commit Suicide allegedly due to exorbitant college fees', 24 January 2016. https://timesofindia.indiatimes.com/india/3-girl-students-allegedly-commit-suicide-in-Tamil-Nadu-due-to-exorbitant-college-fee/articleshow/50704127.cms.

Thiagarajan, Radha. 2004. *Karumuttu Thiagaraja Chettiar, the Textile King*. Chennai: Vanathi Pathippakam.

Thomas, Pradip. 2008. *Strong Religion, Zealous Media: Christian Fundamentalism and Communication in India*. New Delhi: Sage.

Trawick, Margaret. 1990. *Notes on Love in a Tamil Family*. Berkeley, CA: University of California Press.

Tukdeo, Shivali, Chetan Singai, and Aiswarya T. 'New Guidelines aim to make engineering education more inclusive—but may perpetuate inequalities'. *Scroll*, 8 April 2020. https://scroll.in/article/990996/new-guidelines-aim-to-make-engineering-education-more-inclusive-but-may-perpetuate-inequalities.

Uberoi, Patricia. 2002. 'Chicks, Kids and Couples'. *India International Quarterly* 29, No. 3-4 (Winter 2002–Spring 2003): 197–210. https://www.jstor.org/stable/23005826.

Upadhya, Carol. 2007. 'Employment, Exclusion and "Merit" in the Indian IT Industry'. *Economic and Political Weekly* 42, No. 20: 63–68. https://www.jstor.org/stable/4419609.

Upadhya, Carol. 2008. 'Management of Culture and Managing through Culture in the Indian Software Outsourcing Industry'. In *Outpost of the Global Information Economy: Work and Workers in India's Outsourcing Industry*, edited by Carol Upadhya and A. R. Vasavi, 101–135. New Delhi: Routledge.

Upadhya, Carol. 2013. 'Shrink-wrapped souls: managing the self in India's new economy'. In *Enterprise Culture in Neoliberal India: Studies in Youth, Class, Work and Media*, edited by Nandini Gooptu, 93–108. London: Routledge.

Upadhya, Carol. 2016a. 'Engineering equality? Education and im/ mobility in coastal Andhra Pradesh, India'. *Contemporary South Asia* 24, No. 3: 242–256. https://doi.org/10.1080/09584935.2016.1203863.

Upadhya, Carol. 2016b. *Re-engineering India: Work, Capital, and Class and in an Offshore Economy*. Delhi: Oxford University Press.

Upadhya, Carol. 1997. 'Social and cultural strategies of class formation in coastal Andhra Pradesh'. *Contributions to Indian Sociology* 31, No. 2: 169–193. https://doi.org/10.1177/006996697031002001.

Upadhya, Carol and Vasavi, A. R. 2006. *Work, Culture, and Sociality in the IT Indian Industry: A Sociological Study*. Bangalore: NIAS.

Upadhya, Carol and Vasavi, A. R. 2008. 'Introduction'. In *In An Outpost of the Global Information Economy: Work and Workers in India's Outsourcing Industry*, 9–49. New Delhi: Routledge.

Urciuoli, Bonnie. 2008. 'Skills and selves in the new workplace'. *American Ethnologist* 35, No. 2: 211–228. https://www.jstor.org/stable/27667485.

Varma, Pavan K. 2004. *Being Indian: The Truth about why the twenty-first century will be India's*. New Delhi: Penguin.

Vasavi, A. R. 2006. 'Caste Indignities and Subjected Personhoods'. *Economic and Political Weekly* 41, No. 35: 3766–3771. https://www.jstor.org/stable/4418645.

Velraj, R. (dir.). 2014. *Velleyilla Pattathari*. Film. Chennai: Wunderbar Films.

Venkatanarayanan, S. 2019. 'It is high time that Tamil Nadu rationalizes its quota system'. *The Wire*, 17 August 2019. Accessed 12 July 2020. https://thewire.in/uncategorised/tamil-nadu-reservation-quota.

Venkatraman, Shriram. 2017. *Social Media in South India*. Los Angeles: UCL Press.

Venkatesan, S. 2010. '44 institutions to lose deemed varsity status'. *The Hindu*, 18 January 2010. https://www.thehindu.com/news/national/44-institutions-to-lose-deemed-varsitystatus/article13673155.ece#:~:text=The%20Centre%20on%20Monday%20told,did%20not%20meet%20the%20standards.

Vera-Sanso, Penny. 2006. 'Conformity and Contestation: Social Heterogeneity in South Indian Settlements'. In *The Meaning of the Local: Politics of Place in South*

India, edited by Geert De Neve and Henrike Donner, 182–205. Abingdon: Univerity College London Press.

Verma, Rahul. 2020. 'Politicians will pose the biggest challenge to NEP'. *Hindustan Times*, 10 August 2020. Accessed 20 April 2020. https://www.hindustantimes.com/analysis/politicians-will-pose-the-biggest-challenge-to-nep/story-EG3yGujv9Rfg3Xb6WAjp2I.html.

Vijayabaskar, M. and Andrew Wyatt. 2013. 'Economic Change, Politics and Caste: The Case of the Kongu Munnetra Kazhagam'. *Economic and Political Weekly* 48, No. 48: 103–111. https://www.jstor.org/stable/23528935.

Vishwanathan, S. 2005. 'An Admission Reform'. *The Frontline*, 18 June 2005. Accessed 18 February 2017. http://www.thehindu.com/todays-paper/tp-national/tp-tamilnadu/PMK-opposes-move-to-make-JEE-mandatory/article17015971.ece.

Walker, Melanie. 2001. 'Engineering Identities'. *British Journal of Sociology of Education* 22, No. 1: 75–89. https://doi.org/10.1080/01425690020030792.

Walkerdine, Valerie. 1989. 'Femininity as Performance'. *Oxford Review of Education* 15, No. 3: 267–279. https://www.jstor.org/stable/1050418.

Ware, Vron. 1992. 'Moments of Danger: Race, Danger, and Memories of Empire'. *History and Theory* 31, No. 4: 116–137. https://doi.org/10.2307/2505418.

Washbrook, David. 1975. 'The development of caste organisation in south India, 1800–1925'. In *South India: Political Institutions and Political Change*, edited by Christopher J. Baker and David Washbrook, 150–203. Delhi: Macmillan.

Weber, Maximilian. 1947. 'The Nature of Charismatic Authority and its Routinization'. In *Theory of Social and Economic Organization*. Trans. by A. R. Anderson and Talcott Parsons. New York: Free Press.

Wise, Amanda and Selvaraj Velayutham. 2005. 'Moral economies of a translocal village: Obligation and shame among South Indian transnational migrants'. *Global Networks* 5, No. 1: 27–47. https://doi.org/10.1111/j.1471-0374.2005.00106.x.

Wise, Amanda and Selvaraj Velayutham. 2017. 'Transnational Affect and Emotion in Migration Research'. *International Journal of Sociology* 47, No. 2: 116–130. https://doi.org/10.1080/00207659.2017.1300468.

Wiser, William Henricks. 1988. *The Hindu Jajmani System: A Socio-Economic System Interrelating Members of a Hindu Village Community in Services*. New Delhi: Manohar Publishers.

Willis, Paul. 1977. *Learning to Labour: How working class kids get working class jobs*. London: Saxon House.

Wilson, Caroline. 2008. 'The "craze" for medicine and engineering: aspirations and inequalities in Kerala, South India'. Working Paper No. 03/08, Amsterdam School of Social Science Research.

World Bank Gender Data Portal. 2018. 'Share of graduates by field, female (%)'. Online Data Bank. https://genderdata.worldbank.org/indicators/se-ter-grad-fe-zs/?fieldOfStudy=Science%2C%20Technology%2C%20Engineering%20and%20Mathematics%20%28STEM%29&year=2019.

Wyatt, Andrew. 2010. *Party System Change in South India. Political entrepreneurs, patterns and processes*. Abingdon: Routledge.

Wyatt, Andrew K. J. 2020. 'Entrepreneurial Politicians: The Business Careers of Politicians in Tamil Nadu, South India'. Working Paper, University of Bristol.

Zacharias, Paul. 2008. 'The Lakshmi of our times'. *Tehelka*, 18 January 2008. Accessed 6 July 2020. https://web.archive.org/web/20080117204715/http://www.tehelka.com/story_main31.asp?filename=Ne300607The_Lakshmi.asp.

Zacharias, Usha. 2013. 'To Race with the Able? Soft Skills and the Psychologisation of Marginality'. In *Beyond Inclusion: The Practice of Equal Access in Indian Higher Education*, edited by Satish Deshpande and Usha Zacharias, 289–327. New Delhi: Routledge.

Index

For the benefit of digital users, indexed terms that span two pages (e.g., 52–53) may, on occasion, appear on only one of those pages.

Tables are indicated by *t* following the page number

admission
   government quota 9, 41
   management quota 9, 41, 113*t*
affect
   caste as 22, 26–27, 29, 31
   economy of 22, 103–4
Alagappa Chettiaar 54–55
architectural design 1–4, 5–6, 12–13, 119–20
   and dating 193–94
   Dravidian neo-classicism 119–20

becoming professional 109–11, 119, 168, 169–70
Brahmin domination 57–58, 59–60
business models 48–49, 64–65, 76, 87, 97–98, 99, 101–2, 103

caste 27–29, 58, 216
   conflict and violence 25–27, 29, 193
   higher education, and 40, 65–68, 216
   intercaste marriage and violence against 25–26, 187–88, 218–19
   mobility 10–12, 26–27, 72–73, 75, 217–18
   politics 17–18, 192
   Tamil Nadu, in 42
   technical education, and 30–31, 57–58, 59–60
chapterization 42–45
Chettiar, Karumuttu Thiagarajan 88–89
Chettiars, Nattukottai 35, 83–84
Chinna College of Technology 35, 89, 91

class 41
   gender and 195, 219–20
   making 10–12, 73–74, 219, 220–21
   cosmopolitanism 219–20

Dalits 11–12, 25–27, 29–30, 218–19
dating 193–94
D. G. S. Dhinakaran 64
dowry 191
Dravidian politics 65–68, 85–86, 109
dress codes 120–22, 145–46

educational dispositions 18
educational mediators 48–49
edupreneurship 77–78, 87, 96, 99
electoral politics 90, 92–93
employment training 110–11, 116–17, 138, 139, 157–58, 217–18
engineering colleges 5–7
   expansion of 4, 71–73, 75, 98, 187–88
   literature on 7–9
   number of 4–5, 14–15, 75
English language 118, 123–24, 127–28
   difficulty in 124, 131–32
entrepreneurial politicians 65–68, 77–78, 91–96
ethnography 6, 32
exposure 20–21, 223–24

family 16–17, 141–43, 144–45, 165–67
female infanticide 24–25, 181
fieldwork 6–7, 32, 37

INDEX

films
  about higher education/ engineering college 16–17, 132–33
first graduates (from their family) 9–10, 33–34, 112*t*, 112*t*, 116
Foucault, Michel 22
friendships 172, 174–75, 178, 227

gender 38–40, 143, 157, 162, 167, 179–80, 224–25
  aspiration 221–22
  and engineering streams 147–48, 149
  intersections 24–26, 31–32
  rivalries 157–62
  and sports 209–10
gender performativity 127, 169–71, 174–75, 178–79, 199–200, 209, 225–26
gender segregation
  rules of 3, 12–14, 144–47, 162–63, 167, 198–99

habitus 16, 30, 219–20
heterosexual love and desire 184–85, 193–94, 198, 199, 200–7
  and violence 226
higher education
  as a contradictory resource 10–11, 110–11, 118–19
  critique of 17–18, 31, 221
  subjectivities in 104
hostel life 162, 163–64, 167, 174–75

Indian Institutes of Technology 17–18, 30, 70
information technology boom 4, 7–8, 69
  critique of IT culture 133–36, 223–24
  IT culture 133
'institutional' big men 48–49, 76, 78, 87, 88–89, 108, 216–17
intimate aspirations 185–86, 193–94, 195, 205, 213

Jawaharlal Nehru University 17–18
Jesadimai Pangiraj or Jeppiaar 93–96

Kalam, APJ Abdul 19–20
knowledge work 7–8, 19, 71
Kongunadu 32–33, 36–37, 58, 61–62, 71–72, 73–74, 217, 220

Madras Institute of Technology 56–57, 59–60
marriage
  as a cause of disruption of education 181–84
  consequences 191
  desirable marriage 186–87, 188–90, 191–92, 225–26
  runaway 185–86
Maruthur Gopalan Ramachandran (MGR) 65–68, 92
masculinity
  performances of 147, 162–64
  crisis 113–14
mathas (Hindu monastery) 62–64
merit 31, 59–60, 70, 89, 130–31
middle class
  aspiration 15–16, 70–72, 73–74, 119–20, 122, 219–20, 223
  anxieties 12, 223–24
  culture 21, 70–71, 72–74
missing girls 181–83, 185–86, 225
mobility 16–17, 72–73, 227
moral panic 167–68

Naidu, Rangaswamy P.S.G 54–55
Naidu, G.R. Damodaran 55–56, 85, 107–8
Nakarattar. *See* Chettiars, Nattukottai
Nehruvian Development 19–20, 21, 23, 58–60, 61, 70, 87, 107–8
New Education Policy 1986 68
Non-Resident Indian or 'NRI' 33–34, 41, 123, 198, 219–20

patrifocality 14, 223
patronage 79–82, 85, 87, 216–17
personality training 116–17
philanthropy 54–55, 68–69, 76, 80–81, 100–1
privatization of education 49
  as a crisis 4–5, 7–10, 74–75

critique of 17–19, 60–61, 68–69, 72
processes of 49–50, 60, 61–62, 63, 65–68, 75, 87
professional hypergamy 28–29, 189–90, 191
PSG Institute of Technology 47–48
history of 47, 54–56, 82–83, 85
public institutions 54–55, 60–61

Rajam, C. 55, 56, 59–60
Ramadoss, S. 25–27
recruitment 114–19, 139, 218, 220–21
religion
in Tamil Nadu 41–42
religious institutions
and engineering colleges 62–65
reservation 29, 72–73, 151
respectability 148, 171–72, 174–75, 177, 186, 225
risk 16–17, 139, 193–98
Rohith Vemula 17–18, 29
romance 184–85, 193–94, 196–97, 198, 200–7, 226

Salem 32–33
social mobility
immobility 10, 218
strategies for 10–12, 223–24
soft skills 223–24
student politics 17–18
subjectivity 21, 104, 107–8, 136, 184, 199–200, 211–12, 224–25, 227–28

subjectivities of suitability 184, 213–14
suicide 16–18
surveillance 40, 176, 190, 198–99, 226

Tamil
identity and language 123–24, 130–32, 219–20
kinship 190
'Tamil technocracy' 22
technical education
history of 47–50, 51, 55–57, 58–60
technocratic vision 19–21, 23, 58–59, 85, 87
timepass relationship 193–94, 200–6
triple challenge 131–32

unemployment 14–15, 112–14
unemployable 115–16
uniforms 90

Willis, Paul 10, 110–11, 138–39
women
and higher education ownership 82–85, 143
in science and technology 9–10, 13–14, 23, 149, 157
and business 82–85
women's sexuality 224–25
Women's Day 164–67

youth studies 8, 31–32, 76, 215, 216, 227–28